Trends in Logic

Volume 50

TRENDS IN LOGIC
Studia Logica Library

VOLUME 50

The book series Trends in Logic covers essentially the same areas as the journal Studia Logica, that is, contemporary formal logic and its applications and relations to other disciplines. The series aims at publishing monographs and thematically coherent volumes dealing with important developments in logic and presenting significant contributions to logical research.

Volumes of Trends in Logic may range from highly focused studies to presentations that make a subject accessible to a broader scientific community or offer new perspectives for research. The series is open to contributions devoted to topics ranging from algebraic logic, model theory, proof theory, philosophical logic, non-classical logic, and logic in computer science to mathematical linguistics and formal epistemology. This thematic spectrum is also reflected in the editorial board of Trends in Logic. Volumes may be devoted to specific logical systems, particular methods and techniques, fundamental concepts, challenging open problems, different approaches to logical consequence, combinations of logics, classes of algebras or other structures, or interconnections between various logic-related domains. Authors interested in proposing a completed book or a manuscript in progress or in conception can contact either christi.lue@springer.com or one of the Editors of the Series.

More information about this series at http://www.springer.com/series/6645

Leo Esakia
Author

Guram Bezhanishvili · Wesley H. Holliday
Editors

Heyting Algebras

Duality Theory

Springer

Author
Leo Esakia (Deceased)
Tbilisi, Georgia

Editors
Guram Bezhanishvili
Department of Mathematical Sciences
New Mexico State University
Las Cruces, NM, USA

Wesley H. Holliday
Department of Philosophy and Group in Logic and the Methodology of Science
University of California
Berkeley, CA, USA

Translated by
Anton Evseev (Deceased)
Birmingham, UK

ISSN 1572-6126 ISSN 2212-7313 (electronic)
Trends in Logic
ISBN 978-3-030-12095-5 (hardcover) ISBN 978-3-030-12096-2 (eBook)
ISBN 978-3-030-12098-6 (softcover)
https://doi.org/10.1007/978-3-030-12096-2

Library of Congress Control Number: 2018968096

This Springer imprint is published by the registered company Springer Nature Switzerland AG
The registered company address is: Gewerbestrasse 11, 6330 Cham, Switzerland

Leo Esakia (1934–2010)

Foreword

This translation of Leo Esakia's book on Heyting algebras has been in the making for a long time. The book was originally published in 1985 by the Georgian publishing house Metsniereba (Science). It was the first volume of Esakia's planned two volume monograph on Heyting algebras. The book turned out to be popular among Soviet logicians, and Esakia had begun work on the second volume, an outline of which is presented in the Appendix of the first volume. But after the break of the Soviet Union, the publishing house ceased to exist, and the project died with it.

Logicians and mathematicians in the West were aware of the existence of the book, and there were many requests to translate the book into English. But as far as we know, there was no formal contract to translate the book by any Western publishing house.

Hilary Priestley was among the Western mathematicians interested in the book, having received a copy from Anna Romanowska. In 2003, Hilary enlisted a Russian student, Anton Evseev, to translate the book into English. At the time, Anton was an undergraduate studying Mathematics at the University of Oxford. His hand-written translation was not widely circulated, but Hilary mentioned the translation to Mai Gehrke, who in turn mentioned it to Nick Bezhanishvili and David Gabelaia. After Esakia's death in 2010, several tributes were planned: a special issue of *Studia Logica* (Vol. 100, No. 1–2) dedicated to him appeared in 2012, and a special volume of Outstanding Contributions to Logic in his honor appeared in 2014 (*Leo Esakia on Duality in Modal and Intuitionistic Logics*, Springer). In addition, it was decided that the English translation of Esakia's book be edited for publication. Hilary scanned and emailed Anton's hand-written translation to Nick and David. With the help of Mamuka Jibladze, Nick and David used some funds from Esakia's last grant to hire staff from the A. Razmadze Mathematical Institute to type up the translation. The first round of editing of the translation occurred in the summer of 2012 by Guram Bezhanishvili and the second round in the summer of 2017 by Guram and Wesley Holliday.

At last the final product is before your eyes. The initial work by Anton Evseev made a valuable contribution toward bringing Leo Esakia's classic monograph to a wider audience, and we were very grateful when Anton willingly agreed that the translation could form the basis of a version edited for publication. It is with great sadness that we report Anton's untimely death in February 2017 at the age of 33. The mathematical community has been robbed of an exceptionally talented colleague, and we regret that Anton himself will not see this English version of *Heyting Algebras* in print.

Although many of the main results of Esakia's book have already made their way into the mathematical literature, there is no better way to see them developed than through Esakia's concise and lucid presentation. We hope the publication of this translation will make Esakia's intellectual legacy accessible to a wider audience.

Guram Bezhanishvili
Nick Bezhanishvili
David Gabelaia
Mai Gehrke
Wesley H. Holliday
Mamuka Jibladze
Hilary Priestley

Contents

Editors' Note

The original title of Esakia's book, translated into English, was 'Heyting Algebras I. Duality Theory'. As explained in the Foreword, volume II did not materialize, so we have removed 'I' from the title. The planned contents of volume II are discussed in detail in the Appendix.

Esakia was unhappy that a large number of typos and mathematical mistakes were introduced in the Russian version of his book, which he could not correct, as he was never given an opportunity to do a final proofreading. We took the liberty to make the corrections without flagging them, as doing so would distract the reader. We have, however, added specially marked editorial footnotes when we felt that further explanation was in order. In addition, we filled in gaps in some proofs. Two cases are noteworthy. First, we edited proofs at the end of Chap. 2, where we drew from the paper "Scattered and hereditarily irresolvable spaces in modal logic" by Guram Bezhanishvili and Patrick J. Morandi, *Archive for Mathematical Logic*, Vol. 49, 2010, pp. 343–365, as well as further communication between Guram and Patrick. Second, Julia Ilin pointed out a gap in Esakia's original proof of what is now Theorem 5.13, so we replaced it by an algebraic version of the proof from pp. 158–9 of G. Boolos, *The Logic of Provability* (Cambridge University Press), 1993.

Since the original Russian publication of Esakia's book, some of the terminology in the area has changed. In particular, several objects and results are now named after Esakia: e.g., Esakia spaces, Esakia's lemma, the Blok-Esakia theorem, etc. At the end of this note, we provide a table comparing some of Esakia's original terminology with modern terminology. In some cases, we have opted not to give the most direct translation of a Russian term, opting instead for a more natural English substitute (e.g., we use the term 'skeletal Heyting algebra' instead of the more directly translated 'stencil Heyting algebra').

Throughout we have made light changes to notation for readability and consistency. Parentheses are often omitted after functions and inverses, but they are often added around meets and conjunctions. In some cases, we use modern symbols in place of the original symbols (e.g., for sum and product) or adopt modern notational practices (e.g., in the notation for functors in Sect. 3.3 of Chap. 3). However, we retain Esakia's convention that bold typeface signals a definition.

Esakia's original bibliography referred to Russian translations of some English books. We have changed these entries to refer to the original English versions. In addition, we changed some references from Russian to English translations, even where the original work was in Russian.

The original Russian version of Esakia's book had no index. We thank Evgeny Kuznetsov for providing an index for this English translation. Each page number in the index indicates the first page on which a phrase appears in bold typeface with its definition.

Finally, we thank Luca Carai for carefully reading the manuscript and pointing out a number of typos, and Patrick J. Morandi and Julia Ilin for the help noted above.

Esakia's notation	Modern notation
Strict hybrid	Esakia space
Hybrid	Quasi-ordered Esakia space
Strongly isotone map	p-morphism or bounded morphism
Skeletal filter	Open filter or $\Box$-filter
Upper cone	Upset
Lower cone	Downset
Sub $\grave{X}$	$\mathcal{P}(X)$ or $\wp(X)$
Con X	Up(X)
Ω	$\Omega(X)$ or $\mathcal{O}(X)$
In	**Int** or **IPC**
G	**GL**
Grz	**Grz** or **S4.Grz**

Guram Bezhanishvili
Wesley H. Holliday

Introduction

Although many results in the theory of Heyting lattices were discovered by several authors, the subject owes its existence to the work of Alfred Tarski. Due to his seminal work at the end of the 1930s and in the beginning of the 1940s, the theory of Heyting lattices (or Brouwerian lattices, as they were called at the time) became an independent area with its internal problems, playing an important role in modern mathematics. G. Birkhoff, the founder of lattice theory, expressed the following opinion about Heyting lattices at one of the symposia on lattice theory:

> One easily shows that the open subsets of any T_1-space X form a (dually atomic, complete) Brouwerian lattice $L(X)$. Moreover, this lattice determines X up to homeomorphism.
>
> On the other hand, the notion of a Brouwerian lattice was originally invented by Brouwer, and formalized by Heyting, to provide a model for a logical system in which proofs by contradiction are excluded ("intuitionist logic"). Thus, intuitionist logic and a question in topology give rise to exactly the same class of lattices.
>
> Finally, as was shown by Nakayama and Funayama, the (complete, algebraic) "structure lattice" $\Theta(L)$ of all congruence relations on any given lattice L is also a Brouwerian lattice. Here, we have a single technical lattice-theoretic concept, which plays an important role in logic, set theory, and algebra alike! This illustrates the unifying power of lattice theory—and the essential unity of all mathematics. [1, pp. 20–21]

In addition to the quote of Birkhoff, today it is worthwhile to mention the invasion of Heyting algebras into category theory: the subobject classifier of any topos forms a Heyting algebra. Finally, in the theory of continuous Scott lattices, which has many applications, every distributive continuous lattice is a Heyting lattice.

This little book is dedicated to Heyting algebras and closure algebras, which are closely related to Heyting algebras (by an embedding procedure). The author attempted to develop duality theory for Heyting algebras, connecting the remarkable Stone duality with Kripke models. Measures were taken to make the book accessible to those with little knowledge of logic, especially with such "exotic" areas of it as modality and intuitionism. The result is far from being perfect.

It is well known that the class of Heyting lattices, being a special class of distributive lattices containing all Boolean lattices, is situated strictly in between distributive lattices and Boolean lattices. The theory of distributive lattices is covered in many books. For example, in the encyclopedic monograph by Birkhoff [2], in the book by Grätzer [3], where the theory of distributive lattices is discussed especially carefully and completely, and in the book by Skornjakov [4]. The theory of Boolean algebras is covered in great detail, with many examples, in the books by Sikorski [5] and Vladimirov [6]; I would also like to mention an interesting small book by Halmos [7] and an elegant "miniature" by Dwinger [8] (only 60 pages!).

Regrettably, there are no monographs on Heyting lattices except the book by Rasiowa and Sikorski [9], which has certainly played an important role. There is a special chapter in the book, called "Pseudo-Boolean algebras," that is dedicated to Heyting algebras. The book, however, is very metamathematical.

My attempt to present some results on Heyting algebras (and closure algebras) cannot fill this gap. One can hope that the material in the book enriches the chapter "Heyting Algebras" of the very interesting—in both content and presentation—monograph *Distributive Lattices* by Balbes and Dwinger [10]. This may be the only book in which special chapters are dedicated to the algebras of logical "origin": Heyting algebras (Ch. IX), Post algebras (Ch. X), De Morgan algebras and Lukasiewicz algebras (Ch. XI). This opinion, however, may be subjective.

The distribution of content among the chapters of the book is as follows.

Chapter 1, as indicated by its title, contains preliminary material and develops the adopted system of notation. One can find necessary background from universal algebra (Sect. 1.1), category theory (Sect. 1.2), and topology (Sect. 1.3). The content of Sect. 1.4 on ordered sets and clusters is slightly less traditional. The definition of Heyting lattices, with the most important examples, is given in Sect. 1.5.

Chapter 2 discusses the theory of Heyting algebras and closure algebras. The connection with modal systems and intuitionistic logic is made in Sect. 2.3. Important notions of skeletons and skeletal closure algebras are introduced, and the equivalence of the category of Heyting algebras and the category of skeletal closure algebras is proved in Sect. 2.5.

The first section of Chap. 3 introduces the key notion of a hybrid that "cross-breeds" topology (Stone space) and order (Kripke frame). The main properties of hybrids are given in Sect. 3.2. The category of hybrids and hybrid maps is defined in Sect. 3.3. A detailed proof of duality between the category of Heyting algebras and the category of strict hybrids is given in Sect. 3.4. From this follows the representation theorem that an arbitrary Heyting algebra is isomorphic to the ring of cones of a suitable partially ordered set. Finally, Sect. 3.5 is almost entirely dedicated to Grzegorczyk algebras, which have an interesting origin, curious properties, and are close relatives of Heyting lattices. All sections conclude with brief bibliographic notes.

The Appendix contains a very brief survey of the contents of the second part of the monograph entitled "Heyting algebras II. Additional chapters." The author hopes that, despite its brevity, the appendix will allow the reader to have an idea of the content of Part II. The book concludes with a list of references, with no claims of completeness.[1]

In conclusion, I would like to note that the book is based on the special course given by the author in the 1970s at the Faculty of Mechanics and Mathematics of Tbilisi State University. The opportunity to give such a course was kindly offered to me by the head of the Department of Algebra and Geometry and a fellow of the Georgian Academy of Sciences, George Chogoshvili, whose expertise I have been exploiting for a quarter of a century. I consider it a pleasant duty and honor to express publicly my deep gratitude to Chogoshvili.

References

1. Abbott, J. C. (Eds.). (1970). *Trends in lattice theory*. Symposium held at the United States Naval Academy in May of 1966. Van Nostrand Reinhold Mathematical Studies, No. 31. Van Nostrand Reinhold Co., New York.
2. Balbes, R., & Dwinger, Ph. (1974). *Distributive lattices*. Columbia, MO: University of Missouri Press.
3. Birkhoff, G. (1967). *Lattice theory* (3rd ed., Vol. XXV). American Mathematical Society Colloquium Publications, American Mathematical Society, Providence, R.I.
4. Dwinger, Ph. (1961). *Introduction to Boolean algebras*. Hamburger Mathematische Einzelschriften, Heft 40. Würzburg: Physica-Verlag.
5. Grätzer, G. (1978). *General lattice theory*. Basel: Birkhäuser Verlag.
6. Halmos, P. R. (1963). *Lectures on Boolean algebras*. Van Nostrand Mathematical Studies, No. 1. Princeton, N.J.: D. Van Nostrand Co., Inc.
7. Rasiowa, H., & Sikorski, R. (1963). *The mathematics of metamathematics*. Monografie Matematyczne, Tom 41. Warsaw: Państwowe Wydawnictwo Naukowe.
8. Sikorski, R. (1960). *Boolean algebras*. Ergebnisse der Mathematik und ihrer Grenzgebiete, N. F., Heft 25. Berlin: Springer.
9. Skornjakov, L. A. (1977). *Elements of lattice theory* (Translated from the Russian by V. Kumar, Adam Hilger, Ltd., Bristol). Delhi: Hindustan Publishing Corp.
10. Vladimirov, D. A. (1969). *Boolean algebras (Russian)*. Moscow: Izdat. "Nauka".

[1] *Editorial note*: In this edition, references appear after each chapter instead of at the end of the book.

Chapter 1
Preliminary Notions and Necessary Facts

1.1 Universal Algebra

For two sets A, B, let $A \cup B$ (resp. $A \cap B$) denote the **union** (resp. **intersection**) of these sets. Let $\varnothing$ denote the empty set. If $A \cap B = \varnothing$, we say that the sets A, B are **disjoint**. The **difference** of sets A and B, i.e., the set of elements of A that do not belong to B, will be denoted by $A - B$. If f is a map from a set X to a set Y and $A \subseteq X$, $B \subseteq Y$, then $f(A)$ (the f-**image** of A) denotes the set of all elements $f(x)$ where $x \in A$, and $f^{-1}(B)$ (the f-**preimage** of B) denotes the set of all $x \in X$ such that $f(x) \in B$. If f maps X to Y (i.e., $f : X \to Y$), then f is said to be **surjective** if $f(X) = Y$, **injective** if $f(x) = f(y)$ implies $x = y$, and **bijective** if f is surjective and injective. If X is an arbitrary set and $\mathbb{F}$ a family of nonempty subsets of X, then $\mathbb{F}$ is said to be a **partition** of X if $\bigcup\{A : A \in \mathbb{F}\} = X$ and $A \cap A' = \varnothing$ for any $A, A' \in \mathbb{F}$ such that $A \neq A'$.

Let K be an arbitrary class of algebras of a given fixed signature and A_1, A_2 two algebras in the class K. A map $h : A_1 \to A_2$ is said to be a **homomorphism** if h preserves all signature operations of these algebras. A homomorphism h is said to be **surjective** (resp. **injective, bijective**) if h is surjective (resp. injective, bijective) as a map. The terms epimorphism, monomorphism, bimorphism will be used in their categorial meaning (see below). The notions of subalgebra, quotient algebra, homomorphic image, congruence relation, and direct product have their usual meaning. Let K be an arbitrary class of algebras of a given fixed signature. We adopt the following notation:

$\mathbf{I}(K)$ is the class of all isomorphic copies of algebras from the class K;
$\mathbf{H}(K)$ is the class of all homomorphic images of algebras from K;
$\mathbf{S}(K)$ is the class of all subalgebras of algebras from K;
$\mathbf{P}(K)$ is the class of all direct products of nonempty families of algebras belonging to K.

Recall that a hereditary ($\mathbf{S}(K) \subseteq K$), homomorphically ($\mathbf{H}(K) \subseteq K$), and multiplicatively ($\mathbf{P}(K) \subseteq K$) closed class K of algebras is said to be a **variety**. It is known

G. Bezhanishvili and W. H. Holliday (eds.), *Heyting Algebras*,
Trends in Logic 50, https://doi.org/10.1007/978-3-030-12096-2_1

that a class K of algebras is a variety if and only if it is **equationally definable**.[1] For any class K, the class $\mathbf{HSP}(K)$ is the smallest variety containing K. The class $\mathbf{HSP}(K)$ is said to be the **variety generated by** K. If $K = \{A\}$, we write $\mathbf{HSP}(A)$. As Tarski showed, a class K is equationally definable if and only if $K = \mathbf{HSP}(A)$ for some algebra A.

An algebra A is said to be a **subdirect product** of a family of algebras$\{A_j : j \in J\}$ if there exists an injective homomorphism (i.e., an **embedding**) $h : A \to \prod\{A_j : j \in J\}$ of the algebra A into the direct product of the family $\{A_j : j \in J\}$ such that for each $j \in J$, the composition $p_j \circ h$ is surjective, where $p_j : \prod\{A_j : j \in J\} \to A_j$ is the projection map. An algebra A is said to be **subdirectly irreducible** if A being a subdirect product of a family $\{A_j : j \in J\}$ implies that $p_j \circ h$ is an isomorphism for some $j \in J$, where $h : A \to \prod\{A_j : j \in J\}$ is the corresponding embedding.

1.1.1 Theorem *An algebra is subdirectly irreducible if and only if the set of all its nontrivial congruences has a least element.*

1.1.2 Theorem *If K is a variety, then each algebra in K can be decomposed into a subdirect product of subdirectly irreducible algebras from K.*

According to this theorem, each variety is completely determined by its class of subdirectly irreducible algebras. One can 'make' this class even smaller, as the following theorem shows.

1.1.3 Theorem *A variety K is generated by its subdirectly irreducible finitely generated algebras.*

If a variety is generated by its finite algebras, then it is said to be **finitely approximable**. If all finitely generated algebras of a variety are finite, the variety is said to be **locally finite**.

A class K of algebras is said to have the **congruence extension property** if for each subalgebra A_0 of an algebra $A \in K$ and each congruence Θ_0 on A_0, there is a congruence Θ on A such that $\Theta \cap A_0^2 = \Theta_0$. If K is a variety, then K has the congruence extension property if and only if for each embedding $f : A_1 \to A_2$ and each surjective homomorphism $g : A_1 \to A_3$, there exist a surjective homomorphism $h : A_2 \to A_4$ and an embedding $k : A_3 \to A_4$ such that $h \circ f = k \circ g$.

A class K of algebras is said to be **congruence distributive** if the lattice of congruences $\Theta(A)$ of each algebra A in K is distributive.

Bibliographic notes. For details the reader is referred to any monograph on universal algebra, e.g., [4, 9, 12]. But Part II of the monograph [1] is certainly sufficient.

1.2 Categories

A category K_1 is said to be a **subcategory** of a category K_2 if

[1] *Editorial note*: A class K of algebras is equationally definable if there is a set of identities (see Sect. 26 of [9]) such that K is exactly the class of algebras satisfying those identities.

(1) each object in K_1 is an object in K_2;

(2) for arbitrary objects A, B in K_1, each morphism $f : A \to B$ in K_1 is a morphism in K_2.

The subcategory K_1 is said to be **full** if, in addition,

(3) for arbitrary objects A, B in K_1, each morphism $f : A \to B$ in K_2 is a morphism in K_1.

A morphism f in a category K is said to be a **monomorphism** if $f \circ g = f \circ h$ implies $g = h$ for arbitrary morphisms g and h in K. A morphism f is said to be an **epimorphism** if $g \circ f = h \circ f$ implies $g = h$. A morphism f is said to be a **bimorphism** if f is an epimorphism and a monomorphism. Finally, a morphism $f : A \to B$ is said to be an **isomorphism** if there is a morphism $g : B \to A$ such that $g \circ f = \mathbf{1}_A$ and $f \circ g = \mathbf{1}_B$. In this case, we write $g = f^{-1}$ and say that A and B are **isomorphic**. Every isomorphism is a bimorphism. The converse is not true in general. If morphisms in a category K are simply maps, then every injective morphism is a monomorphism and every surjective morphism is an epimorphism.

An object A in a category K is said to be **injective** if for any monomorphism $f : B \to C$ and any morphism $g : B \to A$ there exists a morphism $h : C \to A$ such that $h \circ f = g$. An object A in K is said to be **projective** if for any epimorphism $f : C \to B$ and any morphism $g : A \to B$ there is a morphism $h : A \to C$ such that $f \circ h = g$.

Categories of algebras. Let K be a class of algebras. Unless stated otherwise, we will associate with the class K the category K (denoted by the same letter and called an **algebraic category**), in which objects are exactly the algebras of the class K and morphisms are the homomorphisms, i.e., maps preserving the signature operations. It is important to note that if K is an algebraic category, then the notions of algebraic and categorical isomorphisms coincide.

Equational categories. An algebraic category K is said to be **equational** if its class of objects forms a variety. If K is an equational category, then monomorphisms coincide with injective homomorphisms, and a direct product of algebras is their product in the categorical sense, but not every epimorphism is surjective.

1.2.1 Definition Let K be an equational category and A an object in K. A subalgebra A_0 of the algebra A is said to be **dense** in A if the inclusion map $h : A_0 \to A$ is an epimorphism.

A subalgebra A_0 of A is dense if and only if for any two morphisms $f, g : A \to B$, if the restrictions of f, g to A_0 coincide, then $f = g$. A category is said to be **balanced** if each bimorphism is an isomorphism. If K is an equational category, then the following three conditions are equivalent:

(1) K is balanced.
(2) Each epimorphism is surjective.
(3) Each dense subalgebra A_0 of an algebra $A \in K$ coincides with A.

The property of being balanced is rather strong: even equational categories are often not balanced, and being balanced implies strong additional properties.

An equational category K is said to be **amalgamable** if for any monomorphisms $f_1 : A_0 \to A_1$, $f_2 : A_0 \to A_2$, there are monomorphisms $g_1 : A_1 \to B$, $g_2 : A_2 \to B$ such that $g_1 \circ f_1 = g_2 \circ f_2$.

Let K_1 and K_2 be arbitrary categories. A **covariant functor** (resp. **contravariant functor**) $F : K_1 \to K_2$ is an assignment of an object $F(A)$ in K_2 to any object A in K_1 and of a morphism $F(f) : F(A) \to F(B)$ in K_2 (resp. $F(f) : F(B) \to F(A)$) to any morphism $f : A \to B$ in K_1 such that

(1) $F(\mathbf{1}_A) = \mathbf{1}_{F(A)}$ for any object A in K_1;
(2) $F(f \circ g) = F(f) \circ F(g)$ (resp. $F(f \circ g) = F(g) \circ F(f)$) for any morphisms f, g in K_1 for which $f \circ g$ is defined.

1.2.2 Definition

(1) A covariant functor $F : K_1 \to K_2$ is said to be an **equivalence** and the categories K_1 and K_2 are said to be **equivalent** if

  (a) for any object A_2 in K_2, there is an object A_1 in K_1 such that $F(A_1)$ and A_2 are isomorphic;
  (b) for any objects A, B in K_1, the map from the set of all morphisms from A to B to the set of all morphisms from $F(A)$ to $F(B)$ induced by the functor is bijective.

(2) A contravariant functor $F : K_1 \to K_2$ is said to be a **coequivalence** (or **duality**) and K_1, K_2 are said to be **dual** if

  (a) for any object A_2 in K_2, there is an object A_1 in K_1 such that $F(A_1)$ and A_2 are isomorphic;
  (b) for any objects A, B in K_1, the map from the set of all morphisms from A to B to the set of all morphisms from $F(B)$ to $F(A)$ induced by F is bijective.

Bibliographic notes. The reader is referred to any book on category theory. Part III of the monograph [1] is certainly sufficient.

1.3 Topologies

A family $\mathbb{K}$ of subsets of an arbitrary set X is said to be a **ring of sets** (over X) if $A, B \in \mathbb{K}$ imply $A \cup B$, $A \cap B \in \mathbb{K}$. A ring of sets is said to be a **field of sets** if $A \in \mathbb{K}$ implies $-A \in \mathbb{K}$ (i.e., $X - A \in \mathbb{K}$). A ring of sets $\mathbb{K}$ (resp. a field of sets $\mathbb{F}$) is said to be **reduced** if for any $x, y \in X$ with $x \neq y$, there is $A \in \mathbb{K}$ (resp. $A \in \mathbb{F}$) such that $x \in A$ and $y \notin A$ or vice versa. Let $\operatorname{Sub} X$ (resp. $\operatorname{Sub}_\omega X$) denote the field of all subsets of X (resp. the field of all finite subsets of X and their complements).

A **topology** on a set X is a family Ω of subsets of X closed under arbitrary unions and finite intersections. Sets from Ω are said to be **open**.[2] A set X equipped with a topology is said to be a **topological space**. A family $\mathbb{B}$ of open subsets of a space (X, Ω) is said to be a **basis** if any set from Ω is a union of some sets from $\mathbb{B}$. A class $\mathbb{B}_0$ of open sets is said to be a **subbasis** if the family $\mathbb{B}$ consisting of $\varnothing$, X and all finite intersections of elements of $\mathbb{B}_0$ is a basis of X.

A topological space (X, Ω) is said to be **compact** if for any family $\{U_j : j \in J\} \subseteq \Omega$ of open sets such that $X = \bigcup\{U_j : j \in J\}$ there exists a finite set $J_0 \subseteq J$ such that $X = \bigcup\{U_j : j \in J_0\}$. A topological space X is said to be a **Kolmogoroff space** (or a **T_0-space**) if for any two distinct points, one of the points belongs to an open set not containing the other point. A topological space is said to be a **T_1-space** if every singleton $\{x\}$ is a closed set. A topological space X is said to be **Hausdorff** if for any two distinct $x, y \in X$, there exist disjoint open sets U_1 and U_2 such that $x \in U_1$ and $y \in U_2$.

1.3.1 Definition A space X is said to be **zero-dimensional** if the small inductive dimension of X is zero ($\operatorname{ind} X = 0$), i.e., for any point $x \in X$ and any open set U containing x, there is a closed open set A such that $x \in A \subseteq U$.[3]

Note that if a space X is zero-dimensional, then the family of all closed open sets of X is a basis for the space. For the purposes of brevity and convenience, following P. Halmos, we call closed and open sets **clopen**.

1.3.2 Proposition *If X is a compact Hausdorff space and $\mathbb{F}$ is a reduced field of clopen sets of X, then $\mathbb{F}$ coincides with the field of all clopen sets of X.*[4]

1.3.3 Definition A **Stone space** is a compact Hausdorff zero-dimensional space.

Note that every Stone space X is **totally disconnected**, i.e., for any two distinct points $x, y \in X$, there is a partition of X into two open sets U_1 and U_2 such that $x \in U_1$, $y \in U_2$. A space X is said to be **extremally disconnected** if the closure of each open set is open, i.e., if each regular open set is clopen.[5]

Let X be a topological space. Let $\exp X$ denote the family of all nonempty closed subsets of X.

1.3.4 Definition The **exponential topology** (**Vietoris topology**) on $\exp X$ is the topology whose subbasis is the family of all sets of the form $\mathbb{F}_1(U) = \{F \in \exp X : F \cap U \neq \varnothing\}$ and $\mathbb{F}_2(U) = \{F \in \exp X : F \subseteq U\}$, where U is an arbitrary open subset of X.

---

[2] *Editorial note*: As usual, a set is **closed** if its complement is open.

[3] *Editorial note*: For the general definition of small inductive dimension, see, e.g., Sect. 7.1 of R. Engelking, *General Topology*, 2nd ed. (Heldermann-Verlag), 1989.

[4] *Editorial note*: For a proof, see Lemma 1 on p. 74 of [10].

[5] *Editorial note*: Recall that a set is **regular open** if it is equal to the interior of its closure.

1.3.5 Proposition *If X is a Stone space, then so is* $\exp X$.[6]

If X is a Stone space, then the sets of the form $\mathbb{F}_1(A)$ and $\mathbb{F}_2(A)$, where A is a clopen subset of X, form a subbasis of $\exp X$.

A subset U of a topological space X is said to be **dense** if the closure of U coincides with X.

1.3.6 Definition

(1) A map f from a topological space X to a space Y is said to be **continuous** if the f-preimage of each open set in Y is open in X.
(2) A bijective map $f : X \to Y$ is said to be a **homeomorphism**, and the spaces X and Y are said to be **homeomorphic**, if f and f^{-1} are continuous.
(3) A map $f : X \to Y$ is said to be **closed** (resp. **open**) if the f-image of each closed (resp. open) set in X is closed (resp. open) in Y.

Any continuous map from a compact space to a Hausdorff space is closed. If, in addition, f is bijective, then f is a homeomorphism.

Let $F = \{(X_j, \Omega_j) : j \in J\}$ be a nonempty family of topological spaces, $X = \prod\{X_j : j \in J\}$ the direct (cartesian) product of the sets X_j $(j \in J)$, and $p_j : X \to X_j$ the projection of X onto X_j (i.e., $p_j(x) = x(j) \in X_j$ for each $x \in X$). Equip the set X with the topology Ω by defining a subbasis to be the family of all sets $p_j^{-1}(U)$ or, equivalently, all sets of the form $\{x \in X : x(j) \in U\}$ where U is an arbitrary open set in X_j. The topology Ω is said to be the **product topology**, and the space X is said to be the **(topological) product** of the X_j $(j \in J)$. The product topology on X is the weakest topology making all projections continuous.

Bibliographic notes. [3, 11].

1.4 Ordered Sets and Clusters

1.4.1 Definition A binary relation R on a set X is said to be

(1) **reflexive** if $(\forall x)(xRx)$;
(2) **transitive** if $(\forall x, y, z)((xRy \;\&\; yRz) \Rightarrow xRz)$;
(3) **connected** if $(\forall x, y)(xRy \vee yRx)$;
(4) **symmetric** if $(\forall x, y)(xRy \Rightarrow yRx)$;
(5) **anti-symmetric** if $(\forall x, y)((xRy \;\&\; yRx) \Rightarrow x = y)$;
(6) **upward directed** (resp. **downward directed**) if $(\forall x, y)(\exists z)(xRz \;\&\; yRz)$ (resp. $(\forall x, y)(\exists z)(zRx \;\&\; zRy)$).

[6] *Editorial note*: This well-known result is due to E. Michael, "Topologies on spaces of subsets," *Transactions of the American Mathematical Society* 71: 152–182, 1951. For a proof using the same definitions as above, see Lemma 2.12 of C. Kupke, A. Kurz, and Y. Venema, "Stone coalgebras," ILLC Report PP-2003-15.

1.4.2 Definition A pair (X, R) is said to be a **quasi-ordered set** if X is a set and R is a reflexive and transitive relation on X. If, in addition, R is anti-symmetric, then (X, R) is said to be a **partially ordered set**.

1.4.3 Definition A subset A of a quasi-ordered set (X, R) is an **upper** (resp. **lower**) **cone** if $x \in A$ and xRy (resp. yRx) imply $y \in A$.[7]

We will often say 'cone' omitting the word 'upper'. It is also convenient to say 'A is an upper (resp. lower) set' instead of 'A is an upper (resp. lower) cone'.

1.4.4 Proposition *Let (X, R) be a quasi-ordered set and $A \subseteq X$. The following conditions are equivalent:*

(1) *A is an upper cone;*
(2) *$R(x) \subseteq A$ for each $x \in A$, where $R(x) = \{y \in X : xRy\}$;*
(3) *$R(A) \subseteq A$, where $R(A) = \bigcup\{R(x) : x \in A\}$;*
(4) *$A = R(A_0)$ for some $A_0 \subseteq X$.*

It is also obvious that any set of the form $R^{-1}(A)$ is a lower cone, where $R^{-1}(A) = \bigcup\{R^{-1}(x) : x \in A\}$ and $R^{-1}(x) = \{y \in X : yRx\}$. Throughout the following notation will be used interchangeably:

$$xRy, \quad R(x, y), \quad (x, y) \in R, \quad y \in R(x), \quad x \in R^{-1}(y).$$

1.4.5 Definition Upper (resp. lower) cones of the form $R(x)$ (resp. $R^{-1}(x)$), where $x \in X$, will be called **principal**.

The following statement is easy to verify.

1.4.6 Proposition *Let (X, R) be a quasi-ordered set. Then:*

(1) *the intersection (resp. union) of any family of cones is a cone;*
(2) *the set-theoretic complement of a lower (resp. upper) cone is an upper (resp. lower) cone;*
(3) *each upper (resp. lower) cone A is the union of all principal upper (resp. lower) cones contained in A.*

1.4.7 Definition A **cluster** of a quasi-ordered set (X, R) is a set of the form $R(x) \cap R^{-1}(x)$, where $x \in X$.

Obviously each cluster A is a set of mutually connected points, i.e., if $x, y \in A$, then xRy and yRx. A cluster A is said to be **nontrivial** if the set A contains at least two distinct points. Note that the classes $E(x) = \{y \in X : xEy\}$ of the natural partition of X induced by the equivalence relation

$$xEy \Leftrightarrow xRy \;\&\; yRx$$

[7] *Editorial note*: Upper (resp. lower) cones are now commonly called 'upward closed sets' (resp. 'downward closed sets') or simply 'upsets' (resp. 'downsets').

are exactly the clusters of the quasi-ordered set (X, R), and

$$E(x) = R(x) \cap R^{-1}(x) \text{ for } x \in X.$$

The corresponding partially ordered quotient set $(X/E, R_E)$, where

$$E(x) R_E E(y) \Leftrightarrow \exists x' \in E(x)\ \exists y' \in E(y) (x' R y'),$$

is called the **skeleton** of the quasi-ordered set (X, R). Due to the importance of clusters for us, we make a few additional remarks. A quasi-ordered set is partially ordered if and only if it has no non-trivial clusters. Each quasi-ordered set X can be turned into a closely related partially ordered set X/E (the skeleton) by "gluing" each cluster of X to a point. The converse procedure of "blowing up" points of a partially ordered set Y to clusters (i.e., replacing a point $x \in Y$ with a corresponding set A_x) gives a quasi-ordered set X, whose skeleton X/E coincides (up to isomorphism, of course) with Y. Moreover, each quasi-ordered set can be obtained this way. Therefore, the prefix 'quasi' indicates the possibility of existence of non-trivial clusters. The following can be considered as an unwritten law of the theory of ordered sets:

If one adds the prefix "quasi-" to any notion, property, etc., which is applicable to a partially ordered set Y, the resulting notion, property, etc. is applicable to any quasi-ordered set X whose skeleton X/E coincides with Y. For example, the statement "Z is a quasi-chain of X" means that the skeleton of the set Z (i.e., Z/E) is a chain of the skeleton X/E of the quasi-ordered set X (see below).

1.4.8 Definition Let (X, R) be a quasi-ordered set and E an equivalence relation (i.e., a reflexive, transitive, and symmetric relation) on X. Then the **partition of** X **induced by** E is the family of classes $E(x) = \{y \in X : xEy\}$ for $x \in X$. A set A contained in X is said to be E-**saturated** if $A = E(A) = \bigcup\{E(x) : x \in A\}$, i.e., if A is a union of equivalence classes of the partition. Obviously E-saturated sets are exactly the sets of the form $E(B)$, where B is a subset of X.

1.4.9 Definition Let (X, R) be a quasi-ordered set. A point $x \in A$ is said to be a **maximal** (resp. **quasi-maximal**) point of $A \subseteq X$ if for any $y \in A$, from xRy it follows that $x = y$ (resp. yRx). The set of all maximal points of A will be denoted by max A. Note that the notions of maximal and quasi-maximal points coincide in partially ordered sets (X, R).[8]

If the relation R of a quasi-ordered set (X, R) is connected, then (X, R) is said to be a **quasi-linearly ordered set** (or a **quasi-chain**). If, in addition, the relation R is anti-symmetric, then (X, R) is said to be a **linearly ordered set** (or a **chain**). A subset Y of a quasi-ordered set (X, R) is said to be a **quasi-chain** (resp. **chain**) if Y with the inherited order, i.e., (Y, R_Y), is a quasi-chain (resp. chain). We will need the following principle, which is equivalent to the axiom of choice.

---

[8] *Editorial note*: The concept of a **minimal** (resp. **quasi-minimal**) point is defined dually. The set of all minimal points of A will be denoted by min A.

1.4.10 Proposition (Hausdorff's Principle) *Any chain of a partially ordered set is contained in a maximal chain.*

1.4.11 Definition Let (X_i, R_i) be quasi-ordered sets, $i = 1, 2$. A map $f : X_1 \to X_2$ is said to be **strongly isotone**[9] if for any $x \in X_1$, $y \in X_2$,

$$f(x)R_2 y \Leftrightarrow \exists x' \in X_1(xR_1x' \ \&\ f(x') = y),$$

or equivalently,

$$f(x) \in R_2^{-1}(y) \Leftrightarrow R_1(x) \cap f^{-1}(y) \neq \varnothing.$$

Obviously if a map f is strongly isotone, then it is **isotone**, i.e., xR_1y implies $f(x)R_2f(y)$ for each $x, y \in X_1$. The following simple but useful proposition holds.

1.4.12 Proposition *Let (X_i, R_i) be quasi-ordered sets $(i = 1, 2)$ and $f : X_1 \to X_2$ a map. The following conditions are equivalent:*

(1) *f is strongly isotone;*
(2) *the f-image and f-preimage of a cone are cones; more precisely, if A is a cone in (X_1, R_1) (resp. in (X_2, R_2)), then $f(A)$ (resp. $f^{-1}(A)$) is a cone in (X_2, R_2) (resp. in (X_1, R_1));*
(3) *$R_1^{-1}(f^{-1}(y)) = f^{-1}(R_2^{-1}(y))$ for each $y \in X_2$;*
(4) *$R_2(f(x)) = f(R_1(x))$ for each $x \in X_1$.*

Proof $(1 \Rightarrow 2)$ Let $f : X_1 \to X_2$ be a strongly isotone map and A a cone in X_1. We show that $f(A)$ is a cone in X_2. Let $y \in f(A)$, i.e., $f(x) = y$ for some $x \in A$, and let yR_2y'. Since f is strongly isotone, there is x' such that xR_1x' and $f(x') = y'$. As A is a cone, $x' \in A$, and so $f(x') = y' \in f(A)$. Therefore, $f(A)$ is a cone in X_2. Now we suppose that A is a cone in X_2 and show that $f^{-1}(A)$ is a cone in X_1. Let $x \in f^{-1}(A)$, i.e., $f(x) \in A$, and let xR_1x'. We show that $x' \in f^{-1}(A)$, i.e., $f(x') \in A$. Since f is isotone, xR_1x' implies $f(x)R_2f(x')$, and as A is a cone, $f(x') \in A$.

$(2 \Rightarrow 1)$ Suppose (2) holds. Let $y \in R_2(f(x))$. Since $R_1(x)$ is a cone in X_1, $f(R_1(x))$ is a cone in X_2. As $x \in R_1(x)$, $f(x) \in f(R_1(x))$. Since $f(x)R_2y$ and $f(R_1(x))$ is a cone in X_2, $y \in f(R_1(x))$. Therefore, there is $x' \in X_1$ such that xR_1x' and $f(x') = y$. Conversely, let $x' \in X_1$, xR_1x', and $f(x) = y$. We show that $f(x)R_2y$, i.e., that $y \in R_2(f(x))$. Consider the cone $f^{-1}(R_2(f(x)))$. Clearly $x \in f^{-1}(R_2(f(x)))$. Since xR_1x', $x' \in f^{-1}(R_2(f(x)))$, i.e., $f(x') \in R_2(f(x))$. But $f(x') = y$. Consequently, conditions (1) and (2) are equivalent.

$(1 \Leftrightarrow 3)$ The equation $R_1^{-1}(f^{-1}(y)) = f^{-1}(R_2^{-1}(y))$ is equivalent to the following:

$$x \in R_1^{-1}(f^{-1}(y)) \Leftrightarrow x \in f^{-1}(R_2^{-1}(y)).$$

Transform the right-hand side to obtain:

---

[9] *Editorial note*: Strongly isotone maps are now commonly called 'p-morphisms' or 'bounded morphisms'.

$$x \in f^{-1}(R_2^{-1}(y)) \Leftrightarrow f(x) \in R_2^{-1}(y) \Leftrightarrow f(x)R_2y.$$

Transform the left-hand side to obtain:

$$x \in R_1^{-1}(f^{-1}(y)) \Leftrightarrow (\exists x')(xR_1x' \;\&\; f(x') = y).$$

Thus, $f(x)R_2y \Leftrightarrow (\exists x')(xR_1x' \;\&\; f(x') = y)$.

(1 $\Leftrightarrow$ 4) The equation $R_2(f(x)) = f(R_1(x))$ is equivalent to the following:

$$y \in R_2(f(x)) \Leftrightarrow y \in f(R_1(x)),$$

i.e., to $f(x)R_2y \Leftrightarrow (\exists x')(xR_1x' \;\&\; f(x') = y)$. □

1.4.13 Proposition *Let (X, R) be a quasi-ordered set, $(X/E, R_E)$ its skeleton, and $f : X \to X/E$ the canonical map, i.e., $f(x) = E(x)$ for each $x \in X$. Then f is a strongly isotone map.*

Proof Let A be a cone in (X, R). We show that $f(A)$ is a cone in $(X/E, R_E)$. We have $f(A) = \bigcup\{f(x) : x \in A\} = \bigcup\{E(x) : x \in A\}$. Suppose that $E(y) \in f(A)$ and $E(y)R_EE(z)$. Then $y \in A$ and yRz, and since A is a cone, $z \in A$. Therefore, $E(z) \in f(A)$. Thus, $f(A)$ is a cone. Now let B be a cone in $(X/E, R_E)$. We show that $f^{-1}(B)$ is a cone in (X, R). Suppose $y \in f^{-1}(B)$ and yRz. Then $f(y) \in B$. Therefore, $E(y) \in B$. Since yRz, we have $E(y)R_EE(z)$. As B is a cone in $(X/E, R_E)$, we have $E(z) \in B$, so $f(z) \in B$. Thus, $z \in f^{-1}(B)$, and so $f^{-1}(B)$ is a cone in (X, R). Consequently, f is strongly isotone by Proposition 1.4.12(2). □

1.4.14 Definition Let (X_i, R_i) be quasi-ordered sets ($i = 1, 2$). A map $f : X_1 \to X_2$ is said to be an **isomorphism** (and (X_1, R_1), (X_2, R_2) are said to be **isomorphic**) if f is a bijection and f, f^{-1} are isotone maps.

The following proposition is easy to check.

1.4.15 Proposition *A map $f : X_1 \to X_2$ is an isomorphism if and only if f is a bijection satisfying*

$$xR_1y \Leftrightarrow f(x)R_2f(y)$$

for each $x, y \in X_1$.

1.4.16 Proposition *Let $f : X_1 \to X_2$ be a strongly isotone bijection. Then f is an isomorphism.*

Proof Since f is strongly isotone, $xR_1y \Rightarrow f(x)R_2f(y)$. Let $f(x)R_2f(y)$. Then there is x' such that xR_1x' and $f(x') = f(y)$. As f is injective, $x' = y$. Thus, xR_1y. □

1.4.17 Proposition *Let (X_i, R_i) be quasi-ordered sets ($i = 1, 2$) and $f : X_1 \to X_2$ an isotone map. Then* (1) *implies* (2)*, and if f is a bijection, then* (1) *and* (2) *are equivalent, where*:

(1) *f is a strongly isotone map;*
(2) *$f(x)R_2f(y)$ implies the existence of $x', y' \in X_1$ such that $f(x) = f(x')$, $f(y) = f(y')$, and $x'R_1y'$ (i.e., f is a* **strong homomorphism** *of the model (X_1, R_1) to the model (X_2, R_2) in the sense of model theory).*

Proof ($1 \Rightarrow 2$) Let $f(x)R_2f(y)$. By (1), there is y' such that xR_1y' and $f(y) = f(y')$. Therefore, xR_1y', $f(x) = f(x)$, and $f(y) = f(y')$. Thus, (2) holds.

($2 \Rightarrow 1$) Let f be a bijection and suppose $f(x)R_2z$. Since $f^{-1}(z) \neq \varnothing$, there exists $y \in f^{-1}(z)$, i.e., $f(y) = z$ and, by assumption, $f(x)R_2f(y)$. Therefore, by (2), there exist x', y' such that $f(x) = f(x')$, $f(y) = f(y')$, and $x'R_1y'$. As f is also an injection, $x = x'$ and $y = y'$. Thus, xR_1y and $f(y) = z$. □

Bibliographic notes. See the first chapter of [2]. Strongly isotone maps were introduced in [5].[10]

1.5 Heyting Lattices

1.5.1 Definition A lattice $(L, \leq)$ is said to be a lattice with **relative pseudocomplement** if for any elements $a, b \in L$, the set $\{d \in L : a \wedge d \leq b\}$ has a greatest element, which is denoted by $a \to b$.

Note that a lattice with relative pseudocomplement $\to$ always has a greatest element 1, since for each a, we have $a \to a = 1$.

1.5.2 Definition A lattice L with relative pseudocomplement $\to$ is said to be a **Heyting lattice** if L has a least element 0. The element $a \to 0$ of a Heyting lattice $(L, \leq)$ is called the **pseudocomplement** of a and is denoted by $\neg a$.

The existence of a relative pseudocomplement (**implication** for short) yields a number of important properties.

1.5.3 Proposition *A Heyting lattice L is a bounded distributive lattice.*[11]

Proof That L is bounded is obvious. Let $a, b, c \in L$ and let d denote $(a \wedge b) \vee (a \wedge c)$. Since $a \wedge b \leq d$ and $a \wedge c \leq d$, we have $b \leq a \to d$ and $c \leq a \to d$. Therefore, $b \vee c \leq a \to d$, and so

$$a \wedge (b \vee c) \leq a \wedge (a \to d) \leq d = (a \wedge b) \vee (a \wedge c).$$

Thus, $a \wedge (b \wedge c) \leq (a \wedge b) \vee (a \wedge c)$. Since $a \wedge (b \vee c) \geq (a \wedge b) \vee (a \wedge c)$ holds in every lattice, we have $a \wedge (b \vee c) = (a \wedge b) \vee (a \wedge c)$. □

---

[10] *Editorial note*: Also see D.H.J. de Jongh and A.S. Troelstra, "On the connection of partially ordered sets with some pseudo-Boolean algebras," *Indigationes Mathematicae* 28: 317–328, 1966.

[11] *Editorial note*: For the definition of bounded distributive lattices, see Sect. 2.1 of Chap. 2.

Complete Heyting lattices are characterized by the following.

1.5.4 Proposition *A complete lattice L is a Heyting lattice if and only if the following* **infinite distributive law** *holds in L:*

$$a \wedge \Big(\bigvee_{i \in I} a_i\Big) = \bigvee_{i \in I} (a \wedge a_i) \text{ for any set } I.$$

Proof Let L be a complete Heyting lattice. Let b denote the element $\bigvee_{i \in I} (a \wedge a_i)$. Clearly $a \wedge a_i \leq b$ for each $i \in I$. Therefore, for each $i \in I$, $a_i \leq a \to b$, so $\bigvee_{i \in I} a_i \leq a \to b$. Since $a \wedge (a \to b) \leq b$, we have $a \wedge \bigvee_{i \in I} a_i \leq b = \bigvee_{i \in I} (a \wedge a_i)$. Note that $\bigvee_{i \in I} (a \wedge a_i) \leq a \wedge \bigvee_{i \in I} a_i$ holds in every complete lattice. Thus, $a \wedge \bigvee_{i \in I} a_i = \bigvee_{i \in I} (a \wedge a_i)$.

Conversely, let $a, b \in L$. We show that $a \to b = \bigvee \{d \in L : a \wedge d \leq b\}$. By the infinite distributive law, we have

$$a \wedge \bigvee \{d \in L : a \wedge d \leq b\} = \bigvee \{a \wedge d : a \wedge d \leq b\}.$$

Therefore, $a \wedge \bigvee \{d \in L : a \wedge d \leq b\} \leq b$. If $a \wedge d' \leq b$ for some $d' \in L$, then $d' \in \{d \in L : a \wedge d \leq b\}$, and so $d' \leq \bigvee \{d \in L : a \wedge d \leq b\}$. Thus, $\bigvee \{d \in L : a \wedge d \leq b\}$ is the greatest element of the set $\{d \in L : a \wedge d \leq b\}$, and hence $a \to b = \bigvee \{d \in L : a \wedge d \leq b\}$. □

1.5.5 Corollary *A finite distributive lattice is a Heyting lattice.*

1.5.6 Definition A bounded distributive lattice L is said to be a **Boolean lattice** if for each $a \in L$ there exists $b \in L$ such that $a \wedge b = 0$ and $a \vee b = 1$ (b is called the **complement** of a and is denoted by $-a$).[12]

Every Boolean lattice is a Heyting lattice in which $a \to b = -a \vee b$. Conversely, a Heyting lattice is a Boolean lattice if and only if for each $a \in L$, we have $a \vee \neg a = 1$.

1.5.7 Proposition *The following hold in every Heyting lattice:*

(1) $a \to b = 1 \Leftrightarrow a \leq b$;
(2) $a \to a = 1$;
(3) $a \to 1 = 1$;
(4) $0 \to b = 1$;
(5) $a \wedge (a \to b) \leq b$;
(6) $a \wedge (a \to b) = a \wedge b$;
(7) $(a \to b) \wedge b = b$;

[12] *Editorial note*: It is easy to see that if an element of a bounded distributive lattice has a complement, then it has a unique complement. See, e.g., Sect. II.6 of [1].

(8) $(a \to b) \wedge (a \to c) = a \to (b \wedge c)$;
(9) $(a \to c) \wedge (b \to c) = (a \vee b) \to c$;
(10) $(a \to b) \wedge c = ((c \wedge a) \to (c \wedge b)) \wedge c$;
(11) $((a \wedge b) \to a) \wedge c = c$;
(12) $a \leq b \Rightarrow \neg b \leq \neg a$;
(13) $a \leq \neg\neg a$;
(14) $\neg\neg\neg a = \neg a$;
(15) $\neg(a \vee b) = \neg a \wedge \neg b$;
(16) $\neg a \vee \neg b \leq \neg(a \wedge b)$;
(17) $\neg a \vee b \leq a \to b$;
(18) $\neg\neg(a \wedge b) = \neg\neg a \wedge \neg\neg b$;
(19) $a \wedge b = 0 \Leftrightarrow a \leq \neg b$;
(20) $a \wedge \neg a = 0$.

1.5.8 Main examples of Heyting lattices.

(1) **Topology**. The family Ω of all open sets of a topological space X, ordered by inclusion $\subseteq$, is a Heyting lattice (the lattice Ω is complete and the infinite distributive law holds in it: $A \cap \bigcup\{A_j : j \in J\} = \bigcup\{A \cap A_j : j \in J\}$ for each $A \in \Omega$ and $A_j \in \Omega$ where J is an arbitrary set).
(2) **Lattices**. The family $\Theta(L)$ of congruences of each lattice L is a Heyting lattice (the Funayama–Nakayama theorem).
(3) **Distributive lattices**. The family $\mathscr{I}(L)$ of ideals of each distributive lattice is a Heyting lattice with respect to inclusion (the Stone theorem).[13]
(4) **Continuous lattices**. Every continuous distributive lattice is a Heyting lattice. (Let L be a complete lattice. An element $a \in L$ is said to be **compact relative to** $b \in L$ if for each family $\{a_i \in L : i \in I\}$, from $b \leq \bigvee\{a_i \in L : i \in I\}$ it follows that there is a finite set $I_0 \subseteq I$ such that $a \leq \bigvee\{a_i \in L : i \in I_0\}$.[14] A lattice L is said to be **continuous** if every $b \in L$ is the least upper bound of all elements that are compact relative to b.)
(5) **Categories**. The subobject classifier Ω of any topos forms a Heyting lattice. (A **topos** is a cartesian closed bicomplete category in which the subobject classifier exists.)
(6) **Logic**. Propositions ordered by the intuitionistic provability relation form a Heyting lattice.
(7) **Ordered sets**. The family of cones of any partially ordered set is a Heyting lattice (with respect to inclusion).
(8) Any chain with 0 and 1 is a Heyting lattice (for any elements a and b, $a \to b = 1$ if $a \leq b$, and $a \to b = b$ otherwise).
(9) Any Boolean lattice is a Heyting lattice, where $a \to b = -a \vee b$ for any elements a, b.

---

[13] *Editorial note*: An **ideal** of a lattice L is a nonempty subset I of L such that (i) if $a \in I$ and $b \leq a$, then $b \in I$, and (ii) if $a, b \in I$, then $a \vee b \in I$.

[14] *Editorial note*: The relation of a being compact relative to b is more standardly called the **way below** relation (see, e.g., Sect. I.1 of [7]).

(10) Any finite distributive lattice L is a Heyting lattice ($a \to b = \bigvee\{d \in L : a \wedge d \leq b\}$).

Bibliographic notes. The proof of Proposition 1.5.7 can be found in [1, 13]. Examples (1)–(3) of 1.5.8 can be found in [2]. Example (4) of 1.5.8 is in the book [7], which is entirely dedicated to the theory of continuous lattices. For Example (5), see [6] or [8]. The most complete exposition of the relationship between Heyting algebras and intuitionistic logic (Example (6)) is in the book [13]. Examples (7)–(10) can be found in [2].

References

1. Balbes, R., & Dwinger, Ph. (1974). *Distributive lattices*. Columbia: University of Missouri Press.
2. Birkhoff, G. (1967). *Lattice theory* (3rd ed., Vol. XXV). Providence: American Mathematical Society Colloquium Publications, American Mathematical Society.
3. Bourbaki, N. (1966). *Elements of mathematics. General topology*. Paris, Reading: Hermann, Addison-Wesley Publishing Co.
4. Cohn, P. M. (1965). *Universal algebra*. New York: Harper & Row Publishers.
5. Esakia, L. L. (1974). Topological Kripke models. *Soviet Mathematics Doklady*, *15*, 147–151.
6. Freyd, P. (1972). Aspect of topoi. *Bulletin of the Australian Mathematical Society*, *7*, 1–76.
7. Gierz, G., Hofmann, K. H., Keimel, K., Lawson, J. D., Mislove, M. W., & Scott, D. S. (1980). *A compendium of continuous lattices*. Berlin: Springer.
8. Goldblatt, R. I. (1979). *Topoi: The categorical analysis of logic* (Vol. 98). Studies in logic and the foundations of mathematics. Amsterdam: North-Holland Publishing Co.
9. Grätzer, G. (1968). *Universal algebra*. Princeton: D. Van Nostrand Co., Inc.
10. Halmos, P. R. (1963). *Lectures on Boolean algebras* (Vol. 1). Van Nostrand mathematical studies. Princeton: D. Van Nostrand Co., Inc.
11. Kelley, J. L. (1955). *General topology*. Toronto: D. Van Nostrand Company, Inc.
12. Mal'cev, A. I. (1973). In D. Smirnov & M. Taĭclin (Eds.), *Algebraic systems* (Posthumous ed.). Translated from the Russian by B. D. Seckler & A. P. Doohovskoy, Die Grundlehren der mathematischen Wissenschaften (Vol. 192). New York: Springer.
13. Rasiowa, H., & Sikorski, R. (1963). *The mathematics of metamathematics* (Vol. 41). Monografie Matematyczne. Warsaw: Państwowe Wydawnictwo Naukowe.

Chapter 2
Heyting Algebras and Closure Algebras

2.1 Heyting Algebras

2.1.1 A bounded distributive lattice $(L, \leq)$ can be considered as a universal algebra $(L, \wedge, \vee, 0, 1)$ with two binary operations $\vee, \wedge$ and two constants $0, 1$ satisfying the following axioms:

(L1) $(a \vee b) \vee c = a \vee (b \vee c), \quad (a \wedge b) \wedge c = a \wedge (b \wedge c),$
(L2) $a \vee b = b \vee a, \quad a \wedge b = b \wedge a,$
(L3) $a \vee a = a, \quad a \wedge a = a,$
(L4) $a \vee (a \wedge b) = a, \quad a \wedge (a \vee b) = a,$
(L5) $a \wedge (b \vee c) = (a \wedge b) \vee (a \wedge c), \quad a \vee (b \wedge c) = (a \vee b) \wedge (a \vee c),$
(L6) $a \wedge 0 = 0, \quad a \vee 1 = 1.$

Let DL denote the variety (and the corresponding equational category) of bounded distributive lattices.

2.1.2 Definition The **center** of a bounded distributive lattice L is the set of all elements that have a complement.[1] Let $\mathrm{CT}(L)$ denote the center of a lattice $L \in \mathsf{DL}$. The elements of the center are called **central**.

2.1.3 Proposition *The center* $\mathrm{CT}(L)$ *of any lattice* $L \in \mathsf{DL}$ *is a bounded sublattice which is a Boolean lattice.*[2]

2.1.4 Proposition *From the point of view of universal algebra (which we will generally adhere to), Heyting lattices can be defined as algebras* $(H, \vee, \wedge, \to, 0, 1)$ *with three binary operations and two constants satisfying the following axioms:*

(H1) $(H, \vee, \wedge, 0, 1) \in \mathsf{DL}$;
(H2) $a \wedge (a \to b) = a \wedge b$;

[1] *Editorial note*: Recall the definition of a complement from Definition 1.5.6.
[2] *Editorial note*: For a proof, see, e.g., p. 67 of [2].

G. Bezhanishvili and W. H. Holliday (eds.), *Heyting Algebras*,
Trends in Logic 50, https://doi.org/10.1007/978-3-030-12096-2_2

(H3) $a \wedge (b \to c) = a \wedge ((a \wedge b) \to (a \wedge c))$;
(H4) $((a \wedge b) \to a) \wedge c = c$.

We denote the resulting variety (and the corresponding category) by HA. *The algebras* $(H, \vee, \wedge, \to, 0, 1) \in$ HA *are called* **Heyting algebras** (*or* **pseudo-Boolean algebras**).

Proof If H is a Heyting lattice, then its lattice operations $\vee$, $\wedge$ (together with the greatest element 1 and the least element 0) satisfy Eqs. (L1)–(L6) (i.e., condition (H1)) by Proposition 1.5.3, and the implication operation $\to$ satisfies Eqs. (H2)–(H4) by Eqs. (6), (10), (11) of Proposition 1.5.7.

Conversely, suppose $(H, \vee, \wedge, \to, 0, 1)$ is a Heyting algebra. Let $a, b \in H$. By (H2), $a \wedge (a \to b) = a \wedge b \leq b$, i.e., $a \wedge (a \to b) \leq b$. If $a \wedge c \leq b$ for some $c \in H$, then by (H3),

$$c \wedge (a \to b) = c \wedge ((c \wedge a) \to (c \wedge b)) = c \wedge ((a \wedge b \wedge c) \to (b \wedge c)).$$

Applying (H4) yields

$$c \wedge ((a \wedge b \wedge c) \to (b \wedge c)) = c.$$

So $c \wedge (a \to b) = c$, i.e., $c \leq a \to b$. Thus, $a \to b$ is the greatest of the elements c satisfying $a \wedge c \leq b$. □

2.1.5 Proposition *A Heyting algebra* $(H, \vee, \wedge, \to, 0, 1)$ *is a* **Boolean algebra** *if and only if any one of the following conditions holds:*

(B1) *The law of excluded middle:* $a \vee \neg a = 1$;
(B2) *Double negation law:* $\neg\neg a = a$;
(B3) *Peirce's law:* $(a \to b) \to a \leq a$;
(B4) $a \to b \leq \neg a \vee b$.

Boolean algebras are usually considered as algebras $(B, \vee, \wedge, -, 0, 1)$ where $-a$ denotes the complement of an element a. Clearly if $(B, \vee, \wedge, -, 0, 1)$ is a Boolean algebra, then $(B, \vee, \wedge, \to, 0, 1)$ is a Heyting algebra, where $a \to b = -a \vee b$. Therefore, the equational category BA of Boolean algebras is a full subcategory of the category HA of Heyting algebras. We will often denote a Boolean algebra $(B, \vee, \wedge, -, 0, 1)$ by B, without any danger of confusion.

2.1.6 The algebra of open sets of a topological space. Let X be a topological space and Ω the family of all its open sets. For any $A, B \in \Omega$, we have $A \cup B \in \Omega$, $A \cap B \in \Omega$, and $\varnothing, X \in \Omega$, so $(\Omega, \cup, \cap, \varnothing, X)$ is a bounded distributive lattice. For any $A, B \in \Omega$, define $A \to B = -\mathbf{C}(A - B)$, where $A - B = A \cap -B$ and $\mathbf{C}$ is the operator of topological closure in the space X.[3] We show that $\to$ is a relative pseudocomplement.

---

[3]*Editorial note*: The operation $\to$ may be equivalently defined by $A \to B = \mathbf{I}(-A \cup B)$, where $\mathbf{I}$ is the operation of topological interior in the space X.

The condition $A \cap U \subseteq B$ is obviously equivalent to the condition $U \subseteq -(A - B)$, i.e., $(A - B) \subseteq -U$. Since U is an open set, $-U$ is closed. Therefore, $A - B \subseteq -U$ is equivalent to $\mathbf{C}(A - B) \subseteq -U$. Thus, $U \subseteq -\mathbf{C}(A - B)$ is equivalent to $A \cap U \subseteq B$. So the algebra $(\Omega, \cup, \cap, \rightarrow \varnothing, X)$ of open sets of any topological space X is a Heyting algebra.

2.1.7 Definition An element a of a Heyting lattice H is said to be **prime** if:

(1) $a \neq 1$;
(2) $a = b \wedge c$ implies $a = b$ or $a = c$.

Note that condition (2) is equivalent to condition

(2_0) $b \wedge c \leq a$ implies $b \leq a$ or $c \leq a$.

Indeed, suppose (2) holds and $b \wedge c \leq a$. Then $a = (b \wedge c) \vee a = (b \vee a) \wedge (c \vee a)$ and by (2), $a = b \vee a$ or $a = c \vee a$. Therefore, $b \leq a$ or $c \leq a$. Conversely, suppose (2_0) holds and $a = b \wedge c$. Then $a \leq b$, $a \leq c$ and $b \wedge c \leq a$. Applying (2_0) yields $a = b$ or $a = c$.[4]

2.1.8 Proposition *A Heyting algebra H is isomorphic to the lattice Ω of open sets of a suitable topological space X if and only if:*

(1) *H is a complete lattice;*
(2) *every element a is a greatest lower bound of prime elements of H.*

Bibliographic notes. See Chap. IX of [1] or Chap. IV of [19]. For Boolean algebras, see [6, 20]. For 2.1.6, see [19]. Proposition 2.1.8 appears in [5].

2.2 Closure Algebras

2.2.1 Definition An algebra $(B, \vee, \wedge, -, 0, 1, \mathbf{C})$, or $(B, \mathbf{C})$ for short, is said to be a **closure algebra** if $(B, \vee, \wedge, -, 0, 1)$ is a Boolean algebra and the closure operator $\mathbf{C}$ satisfies Kuratowski's axioms:

(C1) $a \leq \mathbf{C}a$;
(C2) $\mathbf{CC}a \leq \mathbf{C}a$;
(C3) $\mathbf{C}(a \vee b) = \mathbf{C}a \vee \mathbf{C}b$;
(C4) $\mathbf{C}0 = 0$

for all $a, b \in B$.

---

[4] *Editorial note*: It is now common to call elements of a lattice satisfying (1) and (2) **meet irreducible** and elements satisfying (1) and (2_0) **meet prime**. The dual concepts are **join irreducible** and **join prime**.

The **interior** I, the dual operator to C, is introduced, as usual, by $\mathsf{I}a = -\mathsf{C}-a$ for each $a \in B$. Obviously $\mathsf{C}a = -\mathsf{I}-a$, so interior algebras $(B, \vee, \wedge, -, 0, 1, \mathsf{I})$ are term equivalent to closure algebras. This is a useful fact.

Let CA denote the class (and the category) of closure algebras. Since every Boolean algebra can be equipped with the "discrete" closure operator C by setting $\mathsf{C}a = a$ for each $a \in B$, the category BA of Boolean algebras can be considered as a full subcategory of CA.

2.2.2 Proposition *Let* $(B, \vee, \wedge, -, 0, 1, \mathsf{C}) \in \mathsf{CA}$. *Then:*

(1) $\mathsf{C}1 = 1$;
(2) $a \leq b \Rightarrow \mathsf{C}a \leq \mathsf{C}b$;
(3) $\mathsf{I}0 = 0$ *and* $\mathsf{I}1 = 1$;
(4) $\mathsf{I}(a \wedge b) = \mathsf{I}a \wedge \mathsf{I}b$;
(5) $\mathsf{I}\mathsf{I}a = \mathsf{I}a$;
(6) $\mathsf{I}a \leq a$;
(7) $a \leq b \Rightarrow \mathsf{I}a \leq \mathsf{I}b$.

2.2.3 Definition An element a of a closure algebra $(B, \vee, \wedge, -, 0, 1, \mathsf{C})$ is said to be **open** (resp. **closed**) if $\mathsf{I}a = a$ (resp. $\mathsf{C}a = a$).

It is easy to see that an element $a \in B$ is open (resp. closed) if and only if $a = \mathsf{I}b$ (resp. $a = \mathsf{C}b$) for some element $b \in B$.

Let $H = H(B, \mathsf{C})$ be the set of all open elements of an algebra $(B, \mathsf{C}) \in \mathsf{CA}$, i.e., let $H = \{\mathsf{I}a : a \in B\}$.

2.2.4 Proposition *The algebra* $(H, \vee, \wedge, \rightarrow, 0, 1)$ *is a Heyting algebra. Moreover:*

(1) $(H, \vee, \wedge, 0, 1)$ *is a sublattice of the Boolean lattice* $(B, \vee, \wedge, 0, 1)$;
(2) *for any* $a, b \in H$,

$$a \rightarrow b = -\mathsf{C}(a \wedge -b) = \mathsf{I}(-a \vee b).$$

Proof (1) is obvious. In order to prove (2), it is sufficient to show that

$$a \wedge d \leq b \Leftrightarrow d \leq \mathsf{I}(-a \vee b) \text{ for any } a, b, d \in H.$$

The condition $a \wedge d \leq b$ is equivalent to $d \leq -a \vee b$. Sine d is an open element, $d \leq -a \vee b$ implies $d = \mathsf{I}d \leq \mathsf{I}(-a \vee b)$ (by Proposition 2.2.2(7)). Since $\mathsf{I}(-a \vee b) \leq -a \vee b$ (Proposition 2.2.2(6)), from $d \leq \mathsf{I}(-a \vee b)$ it follows that $d \leq -a \vee b$. So $a \wedge d \leq b$ is equivalent to $d \leq \mathsf{I}(-a \vee b)$. Note that for any element $a \in H$, $\neg a = (a \rightarrow 0) = -\mathsf{C}a = \mathsf{I}-a$. □

So with each algebra $(B, \mathsf{C}) \in \mathsf{CA}$ is associated the Heyting algebra $H = H(B, \mathsf{C})$ of its open elements, which we call the **skeleton** of the algebra (B, C). Moreover, as we will see below, every Heyting algebra is (up to isomorphism) the skeleton of some closure algebra.

The indicated assignment of objects of category HA to objects of category CA can easily be extended to a (covariant) functor Δ from the category CA to the category HA. Indeed, if $(B, \mathbf{C})$, $(B', \mathbf{C}') \in$ CA, H, H' are their skeletons, and $h : B \to B'$ is a morphism in the category CA, then $\Delta(h) = h\big|_H$ (the restriction of the map h to H) is a morphism $h\big|_H : H \to H'$ in the category HA. Since h is a bounded lattice homomorphism, $h\big|_H$ preserves lattice operations and constants 0, 1; since h preserves the operations of complement and interior, we have

$$h\big|_H(a \to b) = h(\mathbf{I}(-a \vee b)) = \mathbf{I}'(-h(a) \vee h(b)) = h\big|_H(a) \to' h\big|_H(b).$$

Obviously if h is injective, then so is $h\big|_H$. Let h be surjective, and let $a' \in H'$. Then $h^{-1}(a') \neq \varnothing$. If $a \in h^{-1}(a')$, then $\mathbf{I}a \in h^{-1}(a')$. Indeed, $h(\mathbf{I}a) = \mathbf{I}'h(a) = \mathbf{I}'a' = a'$. So $h(\mathbf{I}a) = a'$, and hence $h\big|_H(\mathbf{I}a) = a'$, i.e., $\left(h\big|_H\right)^{-1}(a') \neq \varnothing$.

Let J be a set and for any $j \in J$, let $(B_j, \mathbf{C}_j) \in$ CA and let H_j be its skeleton. Form the direct product $(B, \mathbf{C}) = \prod\left\{(B_j, \mathbf{C}_j) : j \in J\right\}$ of the family $\left\{(B_j, \mathbf{C}_j) : j \in J\right\}$. It is easy to check that the skeleton H of $(B, \mathbf{C})$ coincides with the direct product $\prod\left\{H_j : j \in J\right\}$, where H_j is the skeleton of $(B_j, \mathbf{C}_j)$. Therefore, the following simple but useful theorem holds.

2.2.5 Theorem *The correspondence Δ is a functor from the category* CA *to the category* HA *that satisfies:*

(1) *Δ preserves injections and surjections;*
(2) *Δ commutes with the operation of direct product, i.e.,*

$$\Delta\Big(\prod\left\{(B_j, \mathbf{C}_j) : j \in J\right\}\Big) = \prod\left\{\Delta(B_j, \mathbf{C}_j) : j \in J\right\}.$$

2.2.6 Corollary

(a) *If K is a subvariety of the variety* HA, *then $\Delta^{-1}(K) = \left\{(B, \mathbf{C}) \in \text{CA} : \Delta(B, \mathbf{C}) \in H\right\}$ is a subvariety of the variety* CA.
(b) *If K is a subvariety of the variety* CA, *then $\Delta(K) = \left\{\Delta(B, \mathbf{C}) : (B, \mathbf{C}) \in K\right\}$ is a subvariety of the variety* HA.

2.2.7 Algebras associated with quasi-orders. With every quasi-ordered set (X, R), one can associate the closure algebra $(\mathrm{Sub}\, X, \cup, \cap, -, \varnothing, X, R^{-1})$ and the Heyting algebra $(\mathrm{Con}\, X, \cup, \cap, \to, \varnothing, X)$, where $\mathrm{Sub}\, X$ (resp. $\mathrm{Con}\, X$) is the family of all subsets (resp. all cones) of X, operations $-, \cup, \cap$ are set-theoretic complement, union, intersection, R^{-1} is the operation assigning to a set A its R-preimage $R^{-1}(A)$, and the operation of relative pseudocomplement (implication) $\to$ is defined as follows:

(1) $A_1 \to A_2 = \left\{x \in X : R(x) \cap (A_1 - A_2) = \varnothing\right\}$ for any $A_1, A_2 \in \mathrm{Con}\, X$;

Then the pseudocomplement of a cone A is:

(2) $\neg A = \left\{x \in X : R(x) \cap A = \varnothing\right\}$.

2.2.8 Proposition *If* (X, R) *is a quasi-ordered set, then:*

(1) *the algebra* $(\mathrm{Sub}\,X, \cup, \cap, -, \varnothing, X, R^{-1})$ *is a closure algebra;*
(2) *the algebra* $(\mathrm{Con}\,X, \cup, \cap, \to, \varnothing, X)$ *is a Heyting algebra;*
(3) $\mathrm{Con}\,X$ *is the skeleton of the algebra* $\mathrm{Sub}\,X$.

Proof (1) Obviously $(\mathrm{Sub}\,X, \cup, \cap, -, \varnothing, X)$ is a Boolean algebra. We show that R^{-1} is a closure operator. Axiom (C1) (see Definition 2.1) that $A \subseteq R^{-1}(A)$ follows from the reflexivity of R. The transitivity of R gives $R^{-1}(R^{-1}(A)) \subseteq R^{-1}(A)$ (Axiom (C2)). Since $R^{-1}(A) = \bigcup\{R^{-1}(x) : x \in A\}$, we have (C3). Axiom (C4), that $R^{-1}(\varnothing) = \varnothing$, is obvious.

(2) By Proposition 1.4.6(1), $(\mathrm{Con}\,X, \cup, \cap, \varnothing, X)$ is a bounded distributive lattice. It remains to show that $\to$ is an operation of relative pseudocomplement. Let $A_1, A_2 \in \mathrm{Con}\,X$. First we show that $A_1 \to A_2$ is a cone. Let $x \in A_1 \to A_2$ and xRy. Then $A_1 \cap R(x) \subseteq A_2$ and $y \in R(x)$. Since $R(y) \subseteq R(x)$, we have $A_1 \cap R(y) \subseteq A_2$, so $y \in A_1 \to A_2$. Next we show that $A_1 \cap (A_1 \to A_2) \subseteq A_2$, i.e., that $A_1 \cap \{x \in X : A_1 \cap R(x) \subseteq A_2\} \subseteq A_2$. Let $y \in A_1$ and $A_1 \cap R(y) \subseteq A_2$. Since R is reflexive, $y \in R(y)$, and hence $y \in A_2$. Lastly, let $A_1, A_2, A \in \mathrm{Con}\,X$ and $A_1 \cap A \subseteq A_2$. We show that $A \subseteq A_1 \to A_2$. Let $x \in A$. We must show that $A_1 \cap R(x) \subseteq A_2$. If $z \in A_1 \cap R(x)$, then $z \in A_1$ and $z \in R(x)$. Since $x \in A$ and A is a cone, $z \in A$. Thus, $z \in A_1 \cap A$, and by assumption, $z \in A_2$.

(3) We show that $\mathrm{Con}\,X$ is the skeleton of the algebra $\mathrm{Sub}\,X$. For any $A \subseteq X$, by Proposition 1.4.4, we have $A \in \mathrm{Con}\,X$ if and only if for any $x \in A$, $R(x) \subseteq A$. Consider the set $\{x \in X : R(x) \subseteq A\}$. We have $\{x \in X : R(x) \subseteq A\} = \{x \in X : (\forall y)(xRy \Rightarrow y \in A)\} = -R^{-1}(-A)$. Since R^{-1} is a closure operator, $-R^{-1}(-A) = \{x \in X : R(x) \subseteq A\}$ is the "interior" of the set A. So $A \in \mathrm{Con}\,X$ if and only if $-R^{-1}(-A) = A$. Let $A_1, A_2 \in \mathrm{Con}\,X$. We show that $A_1 \to A_2 = -R^{-1}(A_1 - A_2)$:

$$A_1 \to A_2 = \{x \in X : A_1 \cap R(x) \subseteq A_2\} = -R^{-1}(A_1 \cap -A_2).$$

This completes the proof. □

2.2.9 Definition Let (X, R) be a quasi-ordered set.

(1) A **ring of cones** $\mathbb{K}$ over (X, R) is a ring $\mathbb{K}$ of sets, each a cone, such that for any $A_1, A_2 \in \mathbb{K}$, the set $\{x \in X : R(x) \cap (A_1 - A_2) = \varnothing\}$ belongs to $\mathbb{K}$.
(2) A **closure field** $\mathbb{F}$ over (X, R) is a field $\mathbb{F}$ of subsets of X such that for any $A \in \mathbb{F}$, the set $\{x \in X : R(x) \cap A \neq \varnothing\}$ belongs to $\mathbb{F}$.

Clearly every ring of cones $\mathbb{K}$ over (X, R) forms a Heyting algebra in which the implication is given by

$$A_1 \to A_2 = \{x \in X : R(x) \cap (A_1 - A_2) = \varnothing\}.$$

In Chap. 3 we will prove a representation theorem stating that every Heyting algebra (resp. closure algebra) is isomorphic to a ring of cones (resp. a closure field) over a suitable quasi-ordered set.

2.2.10 Closure algebras associated with topological spaces. Let X be a topological space and **C** the operation of topological closure in this space. Obviously $(\mathrm{Sub}\,X, \cup, \cap, -, \varnothing, X, \mathbf{C})$ is a closure algebra. The algebra obtained in this way will be called the **closure algebra over the topological space** X. Note that the Heyting algebra $(\Omega, \cup, \cap, \to, \varnothing, X)$ of open sets in X is the skeleton of $(\mathrm{Sub}\,X, \cup, \cap, -, \varnothing, X, \mathbf{C})$ (see 2.1.6).

Bibliographic notes. The most complete presentation of closure algebras is given in [19], Chap. III; a very brief one is given in Chap. IX of [1]. A proof of Proposition 2.2.2 can be found in [19]. The subject owes its existence to [17]. Theorem 2.2.5 and its corollary can be found in [10] and in another form in [4, 16], as well as in the interesting monograph [3].

2.3 Modal Systems and Superintuitionistic Logics

As usual, formulae of intuitionistic propositional calculus are constructed from propositional variables using connectives $\vee$, &, $\to$, $\neg$.

2.3.1 Definition A **superintuitionistic logic** is a set of formulae containing all theorems of the intuitionistic propositional calculus **In** and closed under the rules of substitution and modus ponens $(\vdash p, \vdash p \to q \Rightarrow \vdash q)$.[5] Let Λ denote the family of all superintuitionistic logics.

Formulae of the modal propositional system **S4** of Lewis are constructed from propositional variables using the connectives $\vee$, &, $\neg$, and $\Diamond$. Classical implication $\to$ and the necessity operator $\Box$ are defined in a standard way: $p \to q = \neg p \vee q$ and $\Box p = \neg\Diamond\neg p$.

2.3.2 Definition The modal system **S4** is defined as the smallest set of formulae containing all classical tautologies, all axioms $\Box p \to p$, $\Box p \to \Box\Box p$, $\Box(p \to q) \to (\Box p \to \Box q)$, and closed under the rules of substitution, modus ponens, and Gödel's rule $(\vdash p \Rightarrow \vdash \Box p)$.[6] Any set of formulae containing the axioms and closed under the rules mentioned above is said to be a **normal extension** of the system **S4**. Let Σ denote the family of all normal extensions of **S4**; members of Σ will be called **modal systems** and will be denoted by σ, sometimes with an index.

2.3.3 Let λ be a superintuitionistic logic, i.e., $\lambda \in \Lambda$, p a formula, and $(H, \vee, \wedge, \to, 0, 1)$ a Heyting algebra. It is known that p can be considered as an algebraic polynomial if propositional connectives are understood as the corresponding algebraic operations of H and propositional variables are understood as variables

[5] *Editorial note*: The intuitionistic propositional calculus is more commonly denoted by '**Int**' or '**IPC**'. Esakia uses the letters p, q for arbitrary (possibly complex) formulae. He has no special notation for propositional variables. The **rule of substitution** states that if $\vdash p$ and q is obtained from p by uniformly substituting formulae for propositional variables in p, then $\vdash q$.

[6] *Editorial note*: The term 'necessitation rule' is now more common than 'Gödel's rule'.

with range H. With each algebra $H \in \mathsf{HA}$ one can associate the superintuitionistic logic $\lambda = \lambda(H) \in \Lambda$, namely the set of the formulae that (when considered as polynomials) are identically equal to $1 \in H$. The same holds (mutatis mutandis) for modal systems $\sigma \in \Sigma$, so we may adopt the notation $\sigma = \sigma(B, \mathbf{C})$, where $(B, \mathbf{C}) \in \mathsf{CA}$. Moreover, it is well known (and useful for the algebraic approach) that for any logic $\lambda \in \Lambda$ there is a Heyting algebra $H \in \mathsf{HA}$ (for example, the Lindenbaum–Tarski algebra[7] of λ) such that $\lambda = \lambda(H)$. Similarly, $\sigma \in \Sigma$ if and only if $\sigma = \sigma(B, \mathbf{C})$ for some closure algebra $(B, \mathbf{C}) \in \mathsf{CA}$.

If $\lambda = \lambda(H)$, we say that the logic λ is **characterized** by the algebra H or that H is an **exact**, or **adequate**, algebraic model of λ. If $\lambda' \subseteq \lambda(H)$, the algebra H is said to be an algebraic model of $\lambda' \in \Lambda$. This terminology can obviously be extended to classes $K \subseteq \mathsf{HA}$. Namely, the notation $\lambda = \lambda(K)$ means that λ is the set of formulae that are true (identically equal to 1) in any algebra $H \in K$. In this case we say that the class K **defines** (or **characterizes**) the logic $\lambda = \lambda(K)$. Note also that logics $\lambda \in \Lambda$ are in bijective correspondence with varieties K_λ of those Heyting algebras in which all formulae $\rho \in \lambda$ are true. Clearly this terminology is applicable (mutatis mutandis) to modal systems $\sigma \in \Sigma$.

Starting with the fundamental work of McKinsey and Tarski, closure algebras and Heyting algebras are the main tool in the study of modal systems and superintuitionistic logics. Nowadays the theory of algebras associated with modal systems (closure algebras, Halmos' monadic algebras, Grzegorczyk algebras).[8] as well as the theory of algebras associated with superintuitionistic logics (Heyting algebras, Stone lattices, Boolean algebras) form an important and independent area of research with its own internal problems and interesting applications.

2.3.4 Kripke models – the semantics of "possible worlds." In the "post-Kripkean" period, it became possible to formulate the semantics of a number of superintuitionistic logics and modal systems in terms of Kripke models (X, R),[9] where points $x \in X$ of a quasi-ordered set (X, R) are interpreted as "possible worlds," "situations," or "moments of time" and the relation R as the relation of "reachability" (of one world from another), the "ordering of moments of time," etc. Under this approach, the truth values t (truth) and f (falsehood) are assigned not to a sentence p but to a pair (p, x) where $x \in X$. In other words, one associates with each sentence p the set $A_p \subseteq X$ of those "situations" in which p is true. For superintuitionistic logics, A_p is assumed to be a cone of the quasi-ordered set (X, R), and for modal systems, A_p is any subset of X. Clearly the statement 'the modal formula p is true (under all

[7] *Editorial note*: For the definition of the Lindenbaum-Tarski algebra, see, e.g., Sect. VI.10 of [19].

[8] *Editorial note*: Closure algebras provide the algebraic semantics for **S4** and its extensions. The algebraic semantics for arbitrary normal modal logics is based on the more general concept of a Boolean algebra with operators (see, e.g., P. Blackburn, M. de Rijke, and Y. Venema, *Modal Logic* (Cambridge University Press), 2001).

[9] *Editorial note*: The term 'Kripke frame' is now standard. In current terminology, a **Kripke model** is a Kripke frame together with a valuation function assigning to each propositional variable a subset of X, which must be a cone in the case of Kripke models for superintuitionistic logics. Note that the binary relation R in a Kripke frame need not be reflexive and transitive, but these conditions are required to give a semantics for **S4**.

valuations) in the model (X, R)' is equivalent to the statement 'the modal formula p (more precisely, the corresponding polynomial) is identically 1 in the closure algebra $(\mathrm{Sub}\, X, \cup, \cap, -, \varnothing, X, R^{-1})$.' The same holds for formulae of superintuitionistic logics.

So Kripke models, when viewed algebraically, are represented by the closure algebras $(\mathrm{Sub}\, X, \cup, \cap, -, \varnothing, X, R^{-1})$. It is known that a closure algebra $(B, \mathbf{C})$ is isomorphic to the closure algebra of the above kind, i.e., $(B, \vee, \wedge, -, 0, 1, \mathbf{C}) \simeq (\mathrm{Sub}\, X, \cup, \cap, -, \varnothing, X, R^{-1})$, if and only if the Boolean algebra $(B, \vee, \wedge, -, 0, 1)$ is complete and atomic and the closure operator $\mathbf{C}$ is **completely additive**, i.e., $\mathbf{C}(\bigvee_{i \in I} a_i) = \bigvee_{i \in I} \mathbf{C} a_i$ for any set I and $a_i \in B$ for $i \in I$.

Clearly not all closure algebras satisfy these properties. Recent results show that there exist the so-called **unmodellable** modal systems, i.e., systems that have no adequate Kripke models.[10] In other words, the class Σ^* of those modal systems $\sigma \in \Sigma$ for which an adequate algebra $(B, \mathbf{C}) \in \mathsf{CA}$ of the above special kind exists $(\sigma = \sigma(B, \mathbf{C}))$ by no means coincides with the class Σ. Moreover, the cardinality of the difference $\Sigma - \Sigma^*$ is that of continuum. Having eliminated the artificial and non-heuristic nature of the algebraic approach, the semantics of "possible worlds" (Kripke models) has however lost the Lindenbaum–Tarksi kind of guarantee: existence of an adequate model for every system.

A similar situation occurs in the case of superintuitionistic logics. Under the algebraic approach, Kripke models (X, R) are represented by Heyting algebras of the special kind $(\mathrm{Con}\, X, \cup, \cap, \rightarrow, \varnothing, X)$.

2.3.5 Definition An element a of a Heyting algebra H is said to be **strictly compact** if for any set $K \subseteq H$, if $a \leq \bigvee\{b \in H : b \in K\}$ then $a \leq b$ for some $b \in K$. A Heyting algebra H is said to be **strictly compactly generated** if every $a \in H$ is equal to the least upper bound of all strictly compact elements $b \in H$ such that $b \leq a$.[11]

2.3.6 Proposition *A Heyting algebra $(H, \vee, \wedge, \rightarrow, 0, 1)$ is isomorphic to the algebra* Con X *of all cones of a suitable quasi-ordered set (X, R) if and only if the lattice H is complete and strictly compactly generated.*

Proof Clearly every lattice of cones Con X is complete (see Proposition 1.4.6(1)). It is easy to check that a cone A is strictly compact if and only if it is principal, and every cone is the union of principal cones contained in it (see Proposition 1.4.6(3)). So the algebra Con X is strictly compactly generated.

Let $(H, \vee, \wedge, \rightarrow, 0, 1)$ be complete and strictly compactly generated, let X be the set of all strictly compact elements of H, and let $aRb \Leftrightarrow b \leq a$ for any $a, b \in H$. Define a map $f : H \rightarrow \mathrm{Con}\, X$ by $f(a) = R(a) \cap X$. It is easy to check that f is an isomorphism from the Heyting algebra H onto the algebra Con X of cones of the quasi-ordered set (X, R). □

[10] *Editorial note*: Such systems are now commonly called **Kripke incomplete** or **Kripke frame incomplete**.

[11] *Editorial note*: Strictly compact elements are now usually called **completely join prime**, and strictly compactly generated lattices are called **completely join prime generated**.

Clearly not every Heyting algebra possesses the properties mentioned in Proposition 2.3.6. The notion of being unmodellable can be applied to superintuitionistic logics as well. However, we shall not go deep into this subject. To end this section, note that S. Thomason has axiomatically defined a very simple modal system which is not only unmodellable, but has no Kripke models at all!

Bibliographic notes. A fundamental work on the connection between intuitionistic logic and Lewis's modal system **S4** is [18]. A.V. Kuznetsov and his students obtained very interesting and deep results on superintuitionistic logics; we only mention the survey paper [15]. Results connecting modal systems and superintuitionistic logics can be found in [3, 7–9, 13, 16]. For S. Thomason's system "without models," see [21].

2.4 Filters and Congruences

2.4.1 Let L be a bounded distributive lattice, i.e., let $L \in \mathsf{DL}$. A nonempty subset F of L is said to be a **filter** in L if $a \wedge b \in F$ is equivalent to $a \in F$ and $b \in F$. Recall that F is a filter in L if and only if:

(a) $a, b \in F \Rightarrow a \wedge b \in F$;
(b) $a \in F$ and $a \leq b \Rightarrow b \in F$.

A filter is said to be **principal** if it is of the form $[a) = \{b \in L : a \leq b\}$, where $a \in L$. A filter F in L is said to be **proper** if $F \neq L$. A filter F is proper if and only if $0 \notin F$. A filter F is said to be **prime** if it is proper and $a \vee b \in F$ implies $a \in F$ or $b \in F$. A filter F in L is said to be **maximal** if F is proper and for any proper filter F', from $F \subseteq F'$ it follows that $F = F'$. In a distributive lattice, every maximal filter is prime. We recall the following fundamental theorem of M. Stone for distributive lattices.[12]

2.4.2 Prime filter theorem. Let L be a distributive lattice, I an ideal in L, and F a filter in L such that $I \cap F = \varnothing$. Then there exists a prime filter F' in L such that $F \subseteq F'$ and $I \cap F' = \varnothing$.

2.4.3 Corollary *In a distributive lattice L, if $a \nleq b$, then there is a prime filter F in L that contains a and does not contain b.*

2.4.4 Proposition *Let L be a bounded distributive lattice. Then*

(1) *every proper filter F in L is contained in a maximal filter;*
(2) *every filter is an intersection of prime filters;*
(3) *if $F \cup I = L$ and $F \cap I = \varnothing$, then F is a prime filter if and only if I is a prime ideal.*

---

[12] *Editorial note*: See Theorem 6 of M. Stone, "Topological representations of distributive lattices and Brouwerian logics," *Časopis pro pěstování matematiky a fysiky* 67(1): 1–25, 1937.

Let H be a Heyting algebra and F a subset of H.

2.4.5 Proposition *F is a filter in H if and only if:*

(1) $1 \in F$;
(2) $a \in F$ *and* $a \to b \in F$ *imply* $b \in F$.

2.4.6 Proposition *Let F be a filter in H. Then $a \to b \in F$ if and only if $b \in [F, a]$, where $[F, a]$ is the filter generated by F and a.*[13]

Proof If $a \to b \in F$, then since $a \wedge (a \to b) \leq b$, we have $b \in [F, a]$. If $b \in [F, a]$, then there exists $d \in F$ such that $a \wedge d \leq b$. Then $d \leq a \to b$, so $d \in F$ implies $a \to b \in F$. □

2.4.7 Proposition *Let F be a proper filter in H and $\neg a \notin F$. Then the filter $[F, a]$ is proper.*

Proof Follows from $\neg a = a \to 0$ and Proposition 2.4.6. □

For a Heyting algebra H, let $\Theta(H)$ denote the lattice of all congruences and $\mathscr{F}(H)$ the lattice of all filters of H.

2.4.8 Let $F \in \mathscr{F}(H)$. Define a binary relation $\theta(F)$ on H as follows: for any $a, b \in H$, $(a, b) \in \theta(F)$ if and only if there is an element $d \in F$ such that $a \wedge d = b \wedge d$.[14]

2.4.9 Proposition *Let $H \in \mathsf{HA}$ and $F \in \mathscr{F}(H)$. Then:*

(a) *$\theta(F)$ is a congruence of the Heyting algebra H, i.e., $\theta(F) \in \Theta(H)$;*
(b) *the map $\theta : \mathscr{F}(H) \to \Theta(H)$ is an isomorphism of the lattices $\mathscr{F}(H)$ and $\Theta(H)$.*

Recall that the inverse map is defined as follows: if $\theta \in \Theta(H)$, then $F(\theta) = \{a \in H : (a, 1) \in \theta\}$.

2.4.10 Corollary *Let $H, H' \in \mathsf{HA}$ and $h : H \to H'$ be a homomorphism. Then h is injective if and only if $h(a) = 1$ implies $a = 1$.*

2.4.11 Theorem *A Heyting algebra H is subdirectly irreducible (in HA) if and only if there is an element $d \in H$ such that $d \neq 1$ and $a \leq d$ for each $a \in H$ such that $a \neq 1$.*

Proof Let $d \in H$ satisfy the indicated condition. Then the filter $F = \{d, 1\} = [d)$ is obviously the smallest of all filters distinct from the unit filter $[1)$. By Proposition 2.4.9, there is a smallest nontrivial congruence of H. Therefore, by Birkhoff's Theorem (Theorem 1.1.1), H is subdirectly irreducible.

---

[13] *Editorial note*: The **filter** $[F, a]$ **generated by** F **and** a is the smallest filter F' such that $F \subseteq F'$ and $a \in F$. An element b belongs to $[F, a]$ iff there is $d \in F$ such that $a \wedge d \leq b$.

[14] *Editorial note*: Equivalently, $(a, b) \in \theta(F)$ if and only if $a \to b, b \to a \in F$ (see, e.g., Sect. I.13 of [19]).

Let H be subdirectly irreducible. Then (by Birkhoff's Theorem and Proposition 2.4.9) there exists a filter F which is the smallest of all filters distinct from $[1)$. The filter F is principal, i.e., $F = [d)$ for some $d \in H$. Indeed, otherwise for each $a \in F$, there must be $b \in F$ such that $a \neq b$ and $b \leq a$. Therefore, $[a) \subseteq [b) \subseteq F$ and $[a) \neq F$, contradicting the fact that F is the smallest nonunit filter. Now let $a \in H$ and $a \neq 1$. Then since $F = [d)$ is the smallest nonunit filter, $[d) \subseteq [a)$ and so $a \leq d$. □

Let $(B, \mathbf{C})$ be a closure algebra, F a filter of the Boolean algebra B, and H the skeleton of $(B, \mathbf{C})$.

2.4.12 Proposition *The set* $F^0 = \{a \in B : (\exists b)(b \in F \cap H \ \& \ b \leq a)\}$ *satisfies:*

(1) F^0 *is a filter of the Boolean algebra* B*;*
(2) *for each* $a \in F^0$, $\mathbf{I}a \in F^0$*;*
(3) $F^0 \subseteq F$*;*
(4) F^0 *is the greatest of all filters of the Boolean algebra* B *satisfying* (2) *and* (3).

Proof (1) Since $F \cap H$ is a filter of H, $F \cap H$ generates the filter F^0.

(2) Let $a \in F^0$. Then $b \leq a$ for some $b \in F \cap H$. Clearly $\mathbf{I}b = b$ and since $\mathbf{I}b \leq \mathbf{I}a$, we have $\mathbf{I}a \in F^0$.

(3) Obvious.

(4) Let F_1 be a filter of the Boolean algebra B such that $F_1 \subseteq F$ and $a \in F_1$ implies $\mathbf{I}a \in F_1$ for any $a \in B$. We show that $F_1 \subseteq F^0$. Let $a \in F_1$. Then $\mathbf{I}a \in F_1$, and since $F_1 \cap H \subseteq F \cap H$, we have $\mathbf{I}a \in F^0$. As $\mathbf{I}a \leq a$, we conclude that $a \in F^0$. □

2.4.13 Definition A filter F of the Boolean algebra B is said to be a **skeletal filter** of the algebra $(B, \mathbf{C})$ if it coincides with F^0, i.e., $F = F^0$.[15]

2.4.14 Proposition *Let* $(B, \mathbf{C})$ *be a closure algebra and* F *a filter of the Boolean algebra* B*. The following conditions are equivalent:*

(1) *For any* $a \in B$, $a \in F$ *implies* $\mathbf{I}a \in F$*;*
(2) F *is a skeletal filter of the algebra* $(B, \mathbf{C})$.

Proof $(1 \Rightarrow 2)$ Suppose (1) holds. To show that $F \subseteq F^0$, let $a \in F$. Then by (1), $\mathbf{I}a \in F$. Therefore, $\mathbf{I}a \in F \cap H$. Thus, $\mathbf{I}a \in F^0$, and as $\mathbf{I}a \leq a$, we have $a \in F^0$.

$(2 \Rightarrow 1)$ Follows directly from Proposition 2.4.12. □

2.4.15 Proposition *Let* $(B, \mathbf{C})$ *be a closure algebra, and let* F_1 *and* F_2 *be skeletal filters of* $(B, \mathbf{C})$*. Then*

$$F_1 \subseteq F_2 \Leftrightarrow F_1 \cap H \subseteq F_2 \cap H.$$

2.4.16 Notation Let $\mathscr{F}^s(B, \mathbf{C})$ denote the family of all skeletal filters of a closure algebra $(B, \mathbf{C})$.

---

[15] *Editorial note*: Skeletal filters are also commonly called '$\mathbf{I}$-filters' or 'open filters'.

2.4.17 Theorem *The family $\mathscr{F}^s(B, \mathbf{C})$ of all skeletal filters of a closure algebra $(B, \mathbf{C})$ ordered by inclusion is isomorphic to the family of all filters of the skeleton H of $(B, \mathbf{C})$.*

Proof Using Proposition 2.4.12, it is easy to verify that the following maps are isotone and injective: the map assigning to each skeletal filter F of $(B, \mathbf{C})$ the filter $F \cap H$ of H, and the map assigning to each filter G of H the smallest filter F of the Boolean algebra B such that $F \cap H = G$. Consequently, the congruences of $(B, \mathbf{C})$ are in bijective correspondence with the skeletal filters. □

2.4.18 Proposition *Let $(B, \mathbf{C})$ be a closure algebra, H its skeleton, and F a skeletal filter. Then:*

(1) *the filter $F \cap H$ of the algebra H is prime if and only if for any $a, b \in B$, $\mathbf{I}a \vee \mathbf{I}b \in F$ implies $\mathbf{I}a \in F$ or $\mathbf{I}b \in F$;*
(2) *the filter $F \cap H$ is maximal if and only for any $a \in B$, either $a \in F$ or $\mathbf{C}{-}a \in F$.*

Proof (1) Let $F \cap H$ be a prime filter and $\mathbf{I}a \vee \mathbf{I}b \in F$. Suppose that $\mathbf{I}a \notin F$ and $\mathbf{I}b \notin F$. Then $\mathbf{I}a \notin F \cap H$ and $\mathbf{I}b \notin F \cap H$. Since the filter $F \cap H$ is prime, $\mathbf{I}a \vee \mathbf{I}b \notin F \cap H$, so $\mathbf{I}a \vee \mathbf{I}b \notin F$, a contradiction. Conversely, let $a \vee b \in F \cap H$ for $a, b \in H$. Then $\mathbf{I}a = a$, $\mathbf{I}b = b$, and $\mathbf{I}a \vee \mathbf{I}b \in F$. Hence, either $\mathbf{I}a \in F$ or $\mathbf{I}b \in F$, i.e., either $\mathbf{I}a = a \in F \cap H$ or $\mathbf{I}b = b \in F \cap H$.

(2) Let $F \cap H$ be a maximal filter in H and $a \notin F$. Then $\mathbf{I}a \notin F$ and hence $\mathbf{I}a \notin F \cap H$. Since $\mathbf{I}a \notin F \cap H$, the maximality of $F \cap H$ implies that $\mathbf{I}a \wedge b = 0$ for some $b \in F \cap H$. From $\mathbf{I}a \wedge b = 0$ it follows that $b \leq -\mathbf{I}a = \mathbf{C}{-}a$. Since $b \in F$, we conclude that $\mathbf{C}{-}a \in F$. Conversely, let F' be a filter in H, $F \cap H \subseteq F'$ and $F \cap H \neq F'$. Then there exists an element $a \in F'$ such that $a \notin F \cap H$, so $a \notin F$. Therefore, $\mathbf{C}{-}a = -\mathbf{I}a \in F$, and hence $\mathbf{I}{-}\mathbf{I}a \in F$. Thus, $\mathbf{I}{-}\mathbf{I}a \in F \cap H$, and so $\mathbf{I}{-}\mathbf{I}a \in F'$. Since $a = \mathbf{I}a$ and $a \in F'$, we have $\mathbf{I}a \wedge \mathbf{I}{-}\mathbf{I}a \in F'$. But $\mathbf{I}a \wedge \mathbf{I}{-}\mathbf{I}a = \mathbf{I}(\mathbf{I}a \wedge -\mathbf{I}a) = \mathbf{I}0 = 0$. Consequently, $0 \in F'$, i.e., F' is not a proper filter. This proves that the filter $F \cap H$ is maximal. □

It is well known that the lattice of filters of a Boolean algebra is isomorphic to the lattice of its congruences.

2.4.19 Theorem *Let $(B, \mathbf{C})$ be a closure algebra and F a filter of the Boolean algebra B. The following conditions are equivalent:*

(1) *F is a skeletal filter;*
(2) *the congruence $\theta(F)$ of the Boolean algebra B corresponding to the filter F is a congruence of the closure algebra $(B, \mathbf{C})$.*

Proof Let $\theta(F)$ be the congruence of the Boolean algebra B corresponding to the filter F, i.e., $(a, b) \in \theta(F)$ if and only if there exists $d \in F$ such that $a \wedge d = b \wedge d$.

$(1 \Rightarrow 2)$. Let F be a skeletal filter and $(a, b) \in \theta(F)$, i.e., $a \wedge d = b \wedge d$ for some $d \in F$. Then $\mathbf{I}d \in F$ and $\mathbf{I}a \wedge \mathbf{I}d = \mathbf{I}b \wedge \mathbf{I}d$. Therefore, $(\mathbf{I}a, \mathbf{I}b) \in \theta(F)$. Thus, $\theta(F)$ is a congruence of $(B, \mathbf{C})$.

(2 ⇒ 1). Let $\theta(F)$ be a congruence of $(B, \mathbf{C})$, and let $a \in F$, i.e., $(a, 1) \in \theta(F)$. Then $(\mathbf{I}a, \mathbf{I}1) \in \theta(F)$. Therefore, $\mathbf{I}a \wedge d = \mathbf{I}1 \wedge d$ for some $d \in F$. But $\mathbf{I}a \wedge d = \mathbf{I}1 \wedge d = 1 \wedge d = d$, so $d \leq \mathbf{I}a$. Thus, $\mathbf{I}a \in F$. □

Consequently, the congruences of $(B, \mathbf{C})$ are in bijective correspondence with the skeletal filters.

2.4.20 Corollary *The lattice of congruences* $\Theta(B, \mathbf{C})$ *of a closure algebra* $(B, \mathbf{C})$ *is isomorphic to the lattice of congruences of the skeleton of* $(B, \mathbf{C})$.

2.4.21 Corollary *A closure algebra* $(B, \mathbf{C})$ *is subdirectly irreducible* (*in* CA) *if and only if its skeleton is subdirectly irreducible* (*in* HA).

2.4.22 Corollary *The variety of Heyting algebras* HA *and the variety of closure algebras* CA *are congruence-distributive and have the congruence extension property.*

Bibliographic notes. For filters and congruences of Heyting algebras, see Chap. IX of [1] and Chap. IV of [19]. Theorem 2.4.2 is due to M. Stone. For congruences of closure algebras, see Chap. III of [19]. The connection between Heyting algebras and closure algebras has been investigated independently by a number of authors; see [3, 4, 12].

2.5 Skeletal Closure Algebras

2.5.1 Definition Let $(B, \vee, \wedge, 0, 1)$ be a Boolean lattice and $(H, \vee, \wedge, 0, 1)$ a bounded sublattice. The lattice $(H, \vee, \wedge, 0, 1)$ is said to be a **relatively complete sublattice** of the Boolean lattice B if for any $a \in B$, the set $\{b \in H : b \leq a\}$ has a greatest element (in H).[16]

2.5.2 Proposition *Relatively complete sublattices* H *of a Boolean lattice* B *are in bijective correspondence with closure operators* $\mathbf{C}$ *on* B.

Proof Let $\mathbf{C}$ be a closure operator on a Boolean algebra B. Consider the set $H = \{\mathbf{I}a : a \in B\}$ (i.e., the skeleton of the algebra $(B, \mathbf{C})$). It is easy to check that H is a relatively complete sublattice of B, where for $a \in B$, the greatest element of the set $\{b \in H : b \leq a\}$ is $\mathbf{I}a = -\mathbf{C}-a \in H$. Indeed, if $a \in B$, then clearly $a \geq \mathbf{I}a \in H$; and if $b \leq a$ and $b \in H$, then $b = \mathbf{I}b \leq \mathbf{I}a$.

Conversely, let H be a relatively complete sublattice of a Boolean lattice B. Define a map $\mathbf{I} : B \to B$ by setting $\mathbf{I}a$ to be the greatest element of the set $\{b \in H : b \leq a\}$. It is easy to check that $\mathbf{I}$ is an interior operator: $\mathbf{I}1 = 1$ since $1 \in H$; it is immediate from the definition of $\mathbf{I}a$ that $\mathbf{I}a \leq a$; $\mathbf{I}a \leq \mathbf{I}\mathbf{I}a$ follows from transitivity of $\leq$; and since H is a sublattice of B, $\mathbf{I}(a \wedge b) = \mathbf{I}a \wedge \mathbf{I}b$. So $\mathbf{C}a = -\mathbf{I}-a$ is a closure operator on the Boolean algebra B and hence $(B, \mathbf{C}) \in$ CA. It is easy to show that this correspondence is bijective. □

[16] *Editorial note*: For H to be a relatively complete sublattice of B is equivalent to the inclusion of H into B having a right adjoint.

2.5.3 Corollary *Let* $(H, \vee, \wedge, 0, 1)$ *be a relatively complete sublattice of a Boolean lattice* $(B, \vee, \wedge, 0, 1)$. *Then for each* $a, b \in H$, *there exists the implication* $a \to b \in H$, *i.e.,* $(H, \vee, \wedge, \to, 0, 1)$ *is a Heyting algebra.*

Proof Let $a, b \in H$ and $d = -a \vee b \in B$. Since H is relatively complete, the set $\{e \in H : e \leq d\}$ has a greatest element g. It is easy to check that $g = a \to b$. □

2.5.4 Corollary *Let* H *be a relatively complete sublattice of* $B \in \mathsf{BA}$. *Then* H *is the skeleton of the closure algebra* $(B, \mathbf{C})$, *where* $\mathbf{C}a$ *is the least element of the set* $\{-b : b \in H \text{ and } a \wedge b = 0\}$.

Proof This follows from $\mathbf{C}a = -\mathbf{I}-a$ and the fact that $\mathbf{I}a$ is the greatest element of $\{b \in H : b \leq a\}$. □

2.5.5 Remark Let $(H, \vee, \wedge, \to, 0, 1)$ be a Heyting algebra. The class of closure algebras whose skeleton "coincides with" (is isomorphic to) H contains a "privileged" closure algebra $(B(H), \mathbf{C}_H)$ whose Boolean reduct is generated, as a Boolean algebra, by its skeleton H. The algebras $(B(H), \mathbf{C}_H)$, where $H \in \mathsf{HA}$, are said to be **skeletal algebras**, i.e., generated from their skeletons H.

2.5.6 Definition Let $(B, \mathbf{C})$ be a closure algebra and H its skeleton. The algebra $(B, \mathbf{C})$ is said to be a **skeletal algebra** if the smallest Boolean subalgebra B' (of B) containing H coincides with B.

The construction of $(B(H), \mathbf{C}_H)$ from a given skeleton H is well known. We recall one of the methods of the construction.

2.5.7 Construction We start with a more general situation. Let $H \in \mathsf{DL}$. It is known that there exists a Boolean algebra (the **Boolean envelope**) $B(H)$ containing H as a sublattice and generated by H as a Boolean algebra.[17] The algebra $B(H)$ can be constructed as follows. Let X be the set of all prime filters of the lattice H. Define a map $\varphi : H \to \operatorname{Sub} X$, where $\operatorname{Sub} X$ is the Boolean algebra (the field) of all subsets of X, by setting $\varphi(a) = \{p \in X : a \in p\}$ for $a \in H$. Then φ is an injective lattice homomorphism from H into $\operatorname{Sub} X$. For simplicity, identify the lattice H with its φ-image in $\operatorname{Sub} X$, i.e., with the ring of sets $\{\varphi(a) : a \in H\}$. Let $B(H)$ be the smallest Boolean subalgebra of $\operatorname{Sub} X$ that contains H, i.e., let $B(H)$ be the field of sets generated by the ring H. Clearly the smallest and greatest elements 0, 1 of the lattice H and of $B(H)$ are the same (and are equal to $\varnothing$ and X, respectively) and $B(H)$ is determined up to an isomorphism that is the identity on H.

Now let the lattice H be a Heyting algebra, i.e., let $H \in \mathsf{HA}$. Then H is a relatively complete sublattice of its envelope $B(H)$. Indeed, any element $a \in B(H)$ can be represented as $a = \bigwedge\{-a_k \vee b_k : k < n\}$, where $a_k, b_k \in H$.[18] The greatest element (in H) of the set $\{b \in H : b \leq a\}$ is equal to $\bigwedge\{a_k \to b_k : k < n\}$,

[17] *Editorial note*: This Boolean algebra is also called the **free Boolean extension** of H (see, e.g., Sect. V.4 of [1]).

[18] *Editorial note*: For a proof of this fact, see, e.g., Sect. IV.3 of [19].

where $\to$ is the operation of relative pseudocomplement in the Heyting algebra H. Relative completeness of the lattice H allows us (see Proposition 2.5.2) to equip the algebra $B(H)$ with the closure operator $\mathbf{C}_H$, and the skeleton of the resulting closure algebra $(B(H), \mathbf{C}_H)$ coincides with H. So we have a map $\Gamma : \mathsf{HA} \to \mathsf{CA}$ that assigns to an algebra $H \in \mathsf{HA}$ a skeletal algebra $\Gamma(H) = (B(H), \mathbf{C}_H)$, which is constructed canonically from the skeleton H.

Let SA denote the class (and the corresponding category) of skeletal closure algebras. Clearly the category SA is a full subcategory of CA.

2.5.8 Proposition *The map Γ can be extended to a functor from the category HA to the category SA of skeletal closure algebras.*

Proof Let $h: H_1 \to H_2$ be a Heyting homomorphism, where $\Gamma(H_i) = (B(H_i), \mathbf{C}_{H_i})$ $(i = 1, 2)$. Extend the map h to $B(H_1)$ as follows: if $a \in B(H_1)$ and $a = \bigwedge\limits_{k<m} (-a_k \vee b_k)$, where $m \in \omega$ and $a_k, b_k \in H_1$, let $\overline{h}(a) = \bigwedge\limits_{k<m} \big(-h(a_k) \vee h(b_k)\big)$. Then $\overline{h} : B(H_1) \to B(H_2)$ is a well-defined Boolean homomorphism.[19] Let $\mathbf{I}_1 a = -\mathbf{C}_1 - a$. We have

$$\begin{aligned} \overline{h}(\mathbf{I}_1 a) &= \overline{h}\left(\mathbf{I}_1\left(\bigwedge_{k<m}(-a_k \vee b_k)\right)\right) \\ &= \overline{h}\left(\bigwedge_{k<m}(a_k \to b_k)\right) = h\left(\bigwedge_{k<m}(a_k \to b_k)\right) \\ &= \bigwedge_{k<m}(h(a_k) \to h(b_k)) = \mathbf{I}_2\left(\bigwedge_{k<m}(-h(a_k) \vee h(b_k))\right) = \mathbf{I}_2 h(a). \end{aligned}$$

Thus, $\overline{h} = \Gamma(h)$ is a morphism in the category SA.[20] □

2.5.9 Proposition *The functor Γ has the following properties:*

(1) *Γ preserves injections and surjections;*
(2) *A closure algebra $\Gamma(\prod\{H_j : j \in J\})$ can be embedded into the closure algebra $\prod\{\Gamma H_j : j \in J\}$;*
(3) *If J is a finite set, $\Gamma(\prod\{H_j : j \in J\})$ is isomorphic to $\prod\{\Gamma H_j : j \in J\}$.*

Let $\Gamma(K) = \{\Gamma(H) : H \in K\}$, where K is a class of algebras contained in HA.

---

[19] *Editorial note*: To see that this is well-defined, we use the fact (see, e.g., Lemma 1 on p. 97 of [1]) that h extends uniquely to a Boolean homomorphism $\overline{h}$ from $B(H_1)$ to $B(H_2)$, which implies that $\overline{h}(\bigwedge\limits_{k<m} (-a_k \vee b_k)) = \bigwedge\limits_{k<m} \big(-h(a_k) \vee h(b_k)\big)$.

[20] *Editorial note*: The verification that Γ preserves the identity and composition is straightforward.

2.5.10 Corollary

(1) $\Gamma(\mathbf{H}K) = \mathbf{H}(\Gamma K)$;
(2) $\Gamma(\mathbf{S}K) = \mathbf{S}(\Gamma K)$;
(3) $\Gamma(H_1 \times H_2) \simeq \Gamma(H_1) \times \Gamma(H_2)$, *where* $H_1 \times H_2$ *is the direct product of the Heyting algebras* H_1 *and* H_2.

2.5.11 Theorem *The functors* Γ *and* Δ *establish the equivalence of the category* HA *of Heyting algebras and the category* SA *of skeletal algebras.*

Proof One can use Proposition 2.5.8 to verify the conditions for equivalence from Definition 1.2.2. However, we prefer to give a different proof based on the Duality Theorem (see below, Theorems 3.3.4 and 3.4.6). □

We now turn to a classification of the elements in a closure algebra.

2.5.12 Definition Let $(B, \mathbf{C})$ be a closure algebra and $a \in B$. Set:

(1) $\rho a = a \wedge \mathbf{C}(\mathbf{C}a - a)$;
(2) $\pi a = a - \rho a$.

2.5.13 Definition Let $(B, \mathbf{C})$ be a closure algebra. For a natural number $k \in \omega$ and $a \in B$, define:

(1) $\rho^0 a = a, \quad \rho^1 a = \rho a$;
(2) $\rho^{k+1} a = \rho\rho^k a$;
(3) $\pi^k a = a - \rho^k a$.

2.5.14 Lemma *For* $k \in \omega$, *we have* $\rho^k a \geq \rho^{k+1} a$, $\rho^k a \leq a$, *and* $\pi^k a \leq a$.

Proof Easy verification. □

Using the operator ρ, one can classify elements of closure algebras as follows.

2.5.15 Definition Let $(B, \mathbf{C})$ be a closure algebra and $a \in B$.

(1) We say that a has **finite rank** if there exists $m \in \omega$ such that $m \neq 0$ and $\rho^m a = 0$. In this case the **rank** of a is the smallest natural number $k \neq 0$ such that $\rho^k a = 0$.
(2) We call a **cyclic** if $a \neq 0$ and $\rho^k a = \rho^{k+1} a \neq 0$ for some $k \in \omega$.
(3) We say that a has **infinite rank** if a is not cyclic and does not have finite rank.

2.5.16 Proposition

(1) *If* $(B, \mathbf{C})$ *is a closure algebra and* $a \in B$, *then* a *has one (and only one) of the three properties given in Definition 2.5.15. Moreover, if* $a = 0$ *or* $a = 1$, *then the rank of* a *is equal to* 1, *i.e.,* $\rho a = 0$.
(2) *There exists a closure algebra* $(B, \mathbf{C})$ *which has a cyclic element* $a \in B$.
(3) *There exists a closure algebra* $(B, \mathbf{C})$ *in which for any* $k \leq \omega$, *there is an element of rank* k.
(4) *There is a closure algebra* $(B, \mathbf{C})$ *which has an element* a *of infinite rank.*

Proof (1) Obvious.

(2) Let $B = \{0, a, b, 1\}$ be the four-element Boolean algebra and $\mathbf{C}$ be the closure operator on B defined as follows: $\mathbf{C}0 = 0$ and $\mathbf{C}d = 1$ if $d \neq 0$. It is easy to see that b is cyclic.

(3) and (4). Let X be the set of all natural numbers, i.e., $X = \{0, 1, 2, \ldots\}$, and let R be the dual of the natural order on X, i.e., $nRm \Leftrightarrow n \geq m$. Let $B_0 = \mathrm{Sub}_0\, X$ be the field of all finite subsets of X and their complements. Let d be the set of all odd numbers. Finally, let $B = B(X)$ be the smallest field of subsets of X containing d and all sets of the field B_0, i.e., the smallest field generated by the family $B_0 \cup \{d\}$. It is easy to check that elements of B can be represented as $(a \cap d) \cup (b \cap -d)$ for suitable $a, b \in B_0$ and that (B, R^{-1}) is a closure algebra. Let $a_m = \{k : k < m\}$, where m is an even number and $m \neq 0$, and let $b_m = a_m \cap d$. A simple calculation shows that:

(a) $b_m \in B_0 \subseteq B$, and the rank of b_m is $\frac{m}{2}$;

(b) $d \in B$, and d has infinite rank. □

The following useful lemma holds.

2.5.17 Lemma *Let $(B, \mathbf{C})$ be a closure algebra, $a \in B$, and $m \in \omega$. Then*

$$a = \bigvee_{k<m} (\rho^k a - \rho^{k+1} a) \vee \rho^m a.$$

Proof By induction on m. Let $m = 0$. Then $a = \rho^0 a$. Let

$$a = \bigvee_{k<m} (\rho^k a - \rho^{k+1} a) \vee \rho^m a.$$

We show that $a = \bigvee_{k<m+1} (\rho^k a - \rho^{k+1} a) \vee \rho^{m+1} a$. Clearly

$$\begin{aligned}
&\bigvee_{k<m+1} (\rho^k a - \rho^{k+1} a)\\
&= \bigvee_{k<m} (\rho^k a - \rho^{k+1} a) \vee (\rho^m a - \rho^{m+1} a)\\
&= \bigvee_{k<m} (\rho^k a - \rho^{k+1} a) \vee (\rho^m a \wedge -\rho^{m+1} a)\\
&= \left(\bigvee_{k<m} (\rho^k a - \rho^{k+1} a) \vee \rho^m a\right) \wedge \left(\bigvee_{k<m} (\rho^k a - \rho^{k+1} a) \vee -\rho^{m+1} a\right)\\
&= a \wedge \left(\bigvee_{k<m} (\rho^k a - \rho^{k+1} a) \vee -\rho^{m+1} a\right).
\end{aligned}$$

Therefore,

$$
\begin{aligned}
&\bigvee_{k<m+1} (\rho^k a - \rho^{k+1} a) \vee \rho^{m+1} a \\
&= \left(a \wedge \left(\bigvee_{k<m} (\rho^k a - \rho^{k+1} a) \vee -\rho^{m+1} a\right)\right) \vee \rho^{m+1} a \\
&= (a \vee \rho^{m+1} a) \wedge \left(\bigvee_{k<m} (\rho^k a - \rho^{k+1} a) \vee -\rho^{m+1} a \vee \rho^{m+1} a\right) \\
&= (a \vee \rho^{m+1} a) \wedge 1 \\
&= a \vee \rho^{m+1} a \\
&= a
\end{aligned}
$$

because, by Lemma 2.5.14, $\rho^{m+1} a \leq a$. □

2.5.18 Lemma *Let* $(B, \mathbf{C}) \in \mathsf{CA}$ *and* $a \in B$. *Then*

$$\rho^k a - \rho^{k+1} a = \mathbf{C}\rho^k a - \mathbf{C}(\mathbf{C}\rho^k a - \rho^k a).$$

Proof Let $b = \rho^k a$. We show that $b - \rho b = \mathbf{C}b - \mathbf{C}(\mathbf{C}b - b)$, i.e., that

(1) $b - (b \wedge \mathbf{C}(\mathbf{C}b - b)) = \mathbf{C}b - \mathbf{C}(\mathbf{C}b - b)$.

Since

$$
\begin{aligned}
b \wedge -(b \wedge \mathbf{C}(\mathbf{C}b - b)) &= b \wedge \big(-b \vee -\mathbf{C}(\mathbf{C}b - b)\big) \\
&= (b \wedge -b) \vee \big(b \wedge -\mathbf{C}(\mathbf{C}b - b)\big) \\
&= b \wedge -\mathbf{C}(\mathbf{C}b - b),
\end{aligned}
$$

(1) is equivalent to

(2) $b - \mathbf{C}(\mathbf{C}b - b) = \mathbf{C}b - \mathbf{C}(\mathbf{C}b - b)$.

Since $b \leq \mathbf{C}b$, we clearly have $b - \mathbf{C}(\mathbf{C}b - b) \leq \mathbf{C}b - \mathbf{C}(\mathbf{C}b - b)$. It remains to show that

(3) $\mathbf{C}b - \mathbf{C}(\mathbf{C}b - b) \leq b - \mathbf{C}(\mathbf{C}b - b)$.

From $\mathbf{C}b - b \leq \mathbf{C}(\mathbf{C}b - b)$ it follows that

(4) $\mathbf{C}b - \mathbf{C}(\mathbf{C}b - b) \leq b$.

Moreover, since $(\mathbf{C}b - \mathbf{C}(\mathbf{C}b - b)) \wedge \mathbf{C}(\mathbf{C}b - b) = 0$, we have

(5) $\mathbf{C}b - \mathbf{C}(\mathbf{C}b - b) \leq -\mathbf{C}(\mathbf{C}b - b)$.

From (4) and (5) we have

$\mathbf{C}b - \mathbf{C}(\mathbf{C}b - b) \leq b \wedge -\mathbf{C}(\mathbf{C}b - b) = b - \mathbf{C}(\mathbf{C}b - b)$. □

2.5.19 Lemma *Let* $(B, \mathbf{C})$ *be a closure algebra and* $a \in B$. *Then*

$$a - \rho^m a = \bigvee_{k<m} (\rho^k a - \rho^{k+1} a) - \rho^m a.$$

Proof By Lemma 2.5.17, $a = \bigvee_{k<m} (\rho^k a - \rho^{k+1} a) \vee \rho^m a$. Therefore,

$$\begin{aligned} a \wedge -\rho^m a &= \left(\bigvee_{k<m} (\rho^k a - \rho^{k+1} a) \vee \rho^m a \right) \wedge -\rho^m a \\ &= \left(\bigvee_{k<m} (\rho^k a - \rho^{k+1} a) \wedge -\rho^m a \right) \vee (\rho^m a \wedge -\rho^m a) \\ &= \bigvee_{k<m} (\rho^k a - \rho^{k+1} a) - \rho^m a. \end{aligned}$$

□

2.5.20 Theorem *Let* $(B, \mathbf{C})$ *be a closure algebra and* a *an element of* B *of rank* $m \in \omega$. *Then*

$$a = \bigvee_{k<m} (\rho^k a - \rho^{k+1} a).$$

Proof Since the rank of a is equal to m, $\rho^m a = 0$. Now apply Lemma 2.5.17. □

2.5.21 Corollary *Under the conditions of Theorem* 2.5.20,

$$a = \bigvee_{k<m} \big(\mathbf{C}\rho^k a - \mathbf{C}(\mathbf{C}\rho^k a - \rho^k a)\big).$$

Proof Apply Lemma 2.5.18. □

2.5.22 Lemma *Let* $(B, \mathbf{C}) \in \mathsf{CA}$ *and* $a \in B$. *Then:*

(1) $\mathbf{C}\rho^k a \geq \mathbf{C}(\mathbf{C}\rho^k a - \rho^k a)$ $(k < m)$;
(2) $\mathbf{C}(\mathbf{C}\rho^k a - \rho^k a) \geq \mathbf{C}\rho^{k+1} a$ $(k < m - 1)$.

Proof (1) Since $\mathbf{C}\rho^k a - \rho^k a \leq \mathbf{C}\rho^k a$,

$$\mathbf{C}(\mathbf{C}\rho^k a - \rho^k a) \leq \mathbf{C}\rho^k a.$$

(2) Since $\rho^k a \wedge \mathbf{C}(\mathbf{C}\rho^k a - \rho^k a) \leq \mathbf{C}(\mathbf{C}\rho^k a - \rho^k a)$,

$$\mathbf{C}\big(\rho^k a \wedge \mathbf{C}(\mathbf{C}\rho^k a - \rho^k a)\big) \leq \mathbf{C}(\mathbf{C}\rho^k a - \rho^k a).$$

As

$$\rho^k a \wedge \mathbf{C}(\mathbf{C}\rho^k a - \rho^k a) = \rho\rho^k a = \rho^{k+1} a,$$

we obtain $\mathbf{C}\rho^{k+1} a \leq \mathbf{C}(\mathbf{C}\rho^k a - \rho^k a)$. □

2.5.23 Remark For any element $a \in B$ of order m, there is a chain of elements $\rho^0 a \geq \rho^1 a \geq \cdots \geq \rho^k a \geq \cdots \geq \rho^{m-1} a$ such that $a = \bigvee_{k<m} (\rho^k a - \rho^{k+1} a)$ (Theorem 2.5.20). By Corollary 2.5.21 and Lemma 2.5.22, there is a chain $a_0 \geq b_0 \geq a_1 \geq b_1 \geq \cdots \geq a_{m-1} \geq b_{m-1}$ consisting of closed elements of $(B, \mathbf{C})$ such that $a = \bigvee_{k<m} (a_k - b_k)$. If H is the skeleton of $(B, \mathbf{C})$, then since $-a_k, -b_k \in H$, it is clear that the element a belongs to the skeletal subalgebra $(B(H), \mathbf{C}_H)$ of $(B, \mathbf{C})$.

2.5.24 Lemma *Let $(B, \mathbf{C}) \in \mathsf{CA}$ and for each $k < m \in \omega$, let a_k, b_k be closed elements of $(B, \mathbf{C})$ such that $a = \bigvee_{k<m} (a_k - b_k)$. Then there are closed elements c_k, d_k of $(B, \mathbf{C})$ such that $a = \bigvee_{k<m} (c_k - d_k)$ and $c_{k-1} \geq d_{k-1} \geq c_k \geq d_k$.*[21]

Proof By replacing b_k by $a_k \wedge b_k$, we may assume that $a_k \geq b_k$ for each k. For each t with $0 \leq t < m$, set $c_m = 0$ and

$$c_t = \bigvee\{a_{k_0} \wedge \cdots \wedge a_{k_t} \mid 0 \leq k_0 < k_1 < \cdots < k_t < m\},$$

$$e_t = \bigvee\{b_{k_0} \wedge \cdots \wedge b_{k_t} \mid 0 \leq k_0 < k_1 < \cdots < k_t < m\},$$

$$d_t = c_{t+1} \vee e_t.$$

Then $c_0 \geq d_0 \geq c_1 \geq d_1 \geq \cdots \geq c_{m-1} \geq d_{m-1} \geq c_m = 0$ and each c_t and d_t is closed. We claim that:

$$\bigvee_{k<m} (a_k - b_k) = \bigvee_{k<m} (c_k - d_k).$$

We first prove the inequality $\bigvee_{k<m} (a_k - b_k) \leq \bigvee_{k<m} (c_k - d_k)$. Since $c_0 = a_0 \vee \cdots \vee a_{m-1}$, the inequalities between the c_k, d_k show that

$$\bigvee_{k<m} (a_k - b_k) \leq c_0 = (c_0 - d_0) \vee (d_0 - c_1) \vee \cdots \vee (c_{m-1} - d_{m-1}) \vee (d_{m-1} - c_m).$$

Consequently, it suffices to show that $(a_s - b_s) \wedge (d_t - c_{t+1}) = 0$ for each s, t. This in turn is equivalent to $a_s \wedge d_t \leq b_s \vee c_{t+1}$. Since $d_t = c_{t+1} \vee e_t$, this amounts to proving $a_s \wedge e_t \leq b_s \vee c_{t+1}$. For this, by the definition of e_t, it is sufficient to show that $a_s \wedge b_{k_0} \wedge \cdots \wedge b_{k_t} \leq b_s \vee c_{t+1}$ for each $k_0 < \cdots < k_t$. If $s \notin \{k_0, \ldots, k_t\}$, then $a_s \wedge b_{k_0} \wedge \cdots \wedge b_{k_t} \leq a_s \wedge a_{k_0} \wedge \cdots \wedge a_{k_t} \leq c_{t+1}$. If $s \in \{k_0, \ldots, k_t\}$, then $a_s \wedge b_{k_0} \wedge \cdots \wedge b_{k_t} \leq b_s \leq b_s \vee c_{t+1}$. This completes the proof of the inequality.

[21] *Editorial note*: Starting with this lemma, we have made edits to Esakia's proofs in the rest of this chapter as mentioned in the Editors' Note.

We prove the reverse inequality by induction. In the base case, $c_0 = a_0$ and $d_0 = b_0$, so the result is clear. Suppose that $m > 1$ and that the result holds for $m - 1$. Set $c'_{m-1} = 0$ and for $1 \leq t < m - 1$,

$$\begin{aligned} c'_t &= \bigvee \{a_{k_0} \wedge \cdots \wedge a_{k_t} \mid 0 \leq k_0 < k_1 < \cdots < k_t < m-1\}, \\ e'_t &= \bigvee \{b_{k_0} \wedge \cdots \wedge b_{k_t} \mid 0 \leq k_0 < k_1 < \cdots < k_t < m-1\}, \\ d'_t &= c'_{t+1} \vee e'_t. \end{aligned}$$

By induction, we have:

$$\bigvee_{k<m-1} (a_k - b_k) \geq \bigvee_{k<m-1} (c'_k - d'_k).$$

Also,

$$c_t = c'_t \vee (c'_{t-1} \wedge a_{m-1}) \text{ and } e_t = e'_t \vee (e'_{t-1} \wedge b_{m-1}).$$

Therefore,

$$d_t = c'_{t+1} \vee (c'_t \wedge a_{m-1}) \vee e'_t \vee (e'_{t-1} \wedge b_{m-1}).$$

From this we get:

$$d_t = d'_t \vee (c'_t \wedge a_{m-1}) \vee (e'_{t-1} \wedge b_{m-1}).$$

Now,

$$c_{m-1} - d_{m-1} = (a_0 \wedge \cdots \wedge a_{m-1}) - (b_0 \wedge \cdots \wedge b_{m-1}) \leq \bigvee_{k<m} (a_k - b_k).$$

Similarly,

$$\begin{aligned} c_0 - d_0 &= (a_0 \vee \cdots \vee a_{m-1}) - (c_1 \vee (b_0 \vee \cdots \vee b_{m-1})) \\ &\leq (a_0 \vee \cdots \vee a_{m-1}) - (b_0 \vee \cdots \vee b_{m-1}) \\ &\leq \bigvee_{k<m} (a_k - b_k). \end{aligned}$$

If $1 \leq t < m - 1$, then

$$c_t - d_t = (c'_t \vee (c'_{t-1} \wedge a_{m-1})) \wedge \neg d_t = (c'_t \wedge \neg d_t) \vee (c'_{t-1} \wedge a_{m-1} \wedge \neg d_t).$$

Since $d'_t \leq d_t$, we have $c'_t \wedge \neg d_t \leq c'_t \wedge \neg d'_t \leq \bigvee_{k<m-1} (a_k - b_k)$, where the last inequality uses the induction hypothesis. For the other term, we have:

$$\begin{aligned} c'_{t-1} \wedge a_{m-1} \wedge \neg d_t &= c'_{t-1} \wedge a_{m-1} \wedge \neg d'_t \wedge (\neg c'_t \vee \neg a_{m-1}) \wedge (\neg e'_{t-1} \vee \neg b_{m-1}) \\ &= c'_{t-1} \wedge a_{m-1} \wedge \neg d'_t \wedge \neg c'_t \wedge (\neg e'_{t-1} \vee \neg b_{m-1}) \\ &= [c'_{t-1} \wedge a_{m-1} \wedge \neg d'_t \wedge \neg c'_t \wedge \neg e'_{t-1}] \vee \\ &\quad\; [c'_{t-1} \wedge a_{m-1} \wedge \neg d'_t \wedge \neg c'_t \wedge \neg b_{m-1}]. \end{aligned}$$

In the final expression above, the second term is below $a_{m-1} \wedge \neg b_{m-1} = a_{m-1} - b_{m-1}$. It remains to consider the first term. It is equal to

$$\begin{aligned} c'_{t-1} \wedge a_{m-1} \wedge \neg d'_t \wedge \neg(c'_t \vee e'_{t-1}) &= c'_{t-1} \wedge a_{m-1} \wedge \neg d'_t \wedge \neg d'_{t-1} \\ &\le (c'_{t-1} - d'_{t-1}) \le \bigvee_{k<m-1} (a_k - b_k), \end{aligned}$$

where the last inequality uses the induction hypothesis. Thus,

$$\bigvee_{k<m} (c_k - d_k) \le \bigvee_{k<m} (a_k - b_k).$$

Consequently, by induction, the reverse inequality holds. The lemma is therefore proven. □

2.5.25 Lemma *Let* $(B, \mathbf{C}) \in \mathsf{CA}$ *and for each* $k < m \in \omega$, *let* c_k, d_k *be closed elements of* $(B, \mathbf{C})$ *such that* $a = \bigvee_{k<m} (c_k - d_k)$ *and* $c_{k-1} \ge d_{k-1} \ge c_k \ge d_k$. *Then there are closed elements* a_k, b_k *of* $(B, \mathbf{C})$ *such that* $\rho a = \bigvee_{1 \le k<m} (a_k - b_k)$.

Proof By definition, $\rho a = a \wedge \mathbf{C}(\mathbf{C}a - a)$. For simplicity let e_k denote $c_k - d_k$ $(k < m)$. We have:

$$\begin{aligned} a \wedge \mathbf{C}(\mathbf{C}a - a) &= \left(\bigvee_{0 \le k<m} e_k\right) \wedge \mathbf{C}(\mathbf{C}a - a) \\ &= \left(e_0 \vee \bigvee_{1 \le k<m} e_k\right) \wedge \mathbf{C}(\mathbf{C}a - a) \\ &= \big(e_0 \wedge \mathbf{C}(\mathbf{C}a - a)\big) \vee \left(\bigvee_{1 \le k<m} e_k \wedge \mathbf{C}(\mathbf{C}a - a)\right) \end{aligned}$$

Since $c_0 \ge c_1 \ge \cdots \ge c_{m-1}$, we have $a = \bigvee_{k<m} (c_k - d_k) \le \bigvee_{k<m} c_k = c_0$ and hence $\mathbf{C}a \le c_0$ because c_0 is closed. From $c_0 - d_0 \le a$ we obtain $-a \le -c_0 \vee d_0$, which with $c_0 \ge d_0$ implies $\mathbf{C}a - a \le c_0 \wedge (-c_0 \vee d_0) = c_0 \wedge d_0 = d_0$. Then since d_0 is closed, we have $\mathbf{C}(\mathbf{C}a - a) \le d_0$, which implies $e_0 \wedge \mathbf{C}(\mathbf{C}a - a) = 0$. Consequently,

$$
\begin{aligned}
a \wedge \mathbf{C}(\mathbf{C}a - a) &= \left(\bigvee_{1 \le k < m} e_k\right) \wedge \mathbf{C}(\mathbf{C}a - a) \\
&= \bigvee_{1 \le k < m} \big(e_k \wedge \mathbf{C}(\mathbf{C}a - a)\big) \\
&= \bigvee_{1 \le k < m} \big((c_k - d_k) \wedge \mathbf{C}(\mathbf{C}a - a)\big) \\
&= \bigvee_{1 \le k < m} \big((c_k \wedge \mathbf{C}(\mathbf{C}a - a)) - d_k\big).
\end{aligned}
$$

Taking a_k to be $c_k \wedge \mathbf{C}(\mathbf{C}a - a)$ and b_k to be d_k completes the proof. □

2.5.26 Lemma *If a and b are closed elements of $(B, \mathbf{C}) \in \mathsf{CA}$, then $\rho(a - b) = 0$.*

Proof We have:

$$
\begin{aligned}
\rho(a - b) &= (a - b) \wedge \mathbf{C}\big(\mathbf{C}(a - b) - (a - b)\big) \\
&= (a - b) \wedge \mathbf{C}\big(\mathbf{C}(a - b) \wedge -(a \wedge -b)\big) \\
&= (a - b) \wedge \mathbf{C}\big(\mathbf{C}(a - b) \wedge (-a \vee b)\big) \\
&= (a - b) \wedge \mathbf{C}\big((\mathbf{C}(a - b) \wedge -a) \vee (\mathbf{C}(a - b) \wedge b)\big) \\
&= (a - b) \wedge \big(\mathbf{C}(\mathbf{C}(a - b) \wedge -a) \vee \mathbf{C}(\mathbf{C}(a - b) \wedge b)\big) \\
&= \big((a - b) \wedge \mathbf{C}(\mathbf{C}(a - b) \wedge -a)\big) \vee \big((a - b) \wedge \mathbf{C}(\mathbf{C}(a - b) \wedge b)\big).
\end{aligned}
$$

Since $a \wedge -b \le a$, we have $\mathbf{C}(a \wedge -b) \le \mathbf{C}a = a$. From this it follows that $\mathbf{C}(a \wedge -b) \wedge -a = 0$, and hence $(a - b) \wedge \mathbf{C}(\mathbf{C}(a - b) \wedge -a) = 0$. Since $\mathbf{C}(a - b) \wedge b \le b$, we have $\mathbf{C}(\mathbf{C}(a - b) \wedge b) \le \mathbf{C}b = b$. Thus, $\mathbf{C}(\mathbf{C}(a - b) \wedge b) \wedge -b = 0$. Consequently, $(a - b) \wedge \mathbf{C}(\mathbf{C}(a - b) \wedge b) = 0$. □

2.5.27 Proposition *Let $(B, \mathbf{C}) \in \mathsf{CA}$ and $a, a_k, b_k \in B$ for $k < m \in \omega$ such that a_k, b_k are closed and $a = \bigvee_{k<m} (a_k - b_k)$. Then $\rho^m a = 0$.*

Proof By repeated application of Lemmas 2.5.24 and 2.5.25, $\rho^{m-1} a = c - d$ for some closed elements $c, d \in B$. Therefore, $\rho^m a = \rho\rho^{m-1} a = \rho(c - d) = 0$ by Lemma 2.5.26. □

2.5.28 Theorem *Let $(B, \mathbf{C})$ be a closure algebra, H its skeleton, and $B(H)$ the Boolean envelope of the lattice H, i.e., the smallest Boolean subalgebra of B containing H. Then $B(H)$ is closed under the closure operator $\mathbf{C}$, i.e., $(B(H), \mathbf{C}_H)$ is a subalgebra of $(B, \mathbf{C})$.*

Proof It is known that each element $a \in B(H)$ can be represented as $a = \bigvee_{k<m} (a_k - b_k)$, where $m \in \omega$ and $-a_k, -b_k \in H$ for $k < m$.[22] Since $-b_k \to -a_k \in H$

---

[22] *Editorial note*: See, e.g., Sect. IV.3 of [19].

for $k < m$, we have $\bigwedge\limits_{k<m} (-b_k \to -a_k) \in H$. Moreover, using Proposition 2.2.4(2),

$$\begin{aligned}\bigwedge_{k<m}(-b_k \to -a_k) &= \bigwedge_{k<m} \mathbf{I}(b_k \vee -a_k) \\ &= \bigwedge_{k<m} \mathbf{I}-(-b_k \wedge a_k) \\ &= -\bigvee_{k<m} \mathbf{C}(a_k - b_k) \\ &= -\mathbf{C}a.\end{aligned}$$

Therefore, $-\mathbf{C}a \in H$. Since $-1 \in H$, we have $\mathbf{C}a = 1 - (-\mathbf{C}a)$. Thus, $\mathbf{C}a \in B(H)$. □

2.5.29 Remark As a consequence of Lemma 2.5.24, without loss of generality, each $a \in B(H)$ can be represented as $a = \bigvee\limits_{k<m} (a'_k - b'_k)$, where $-a'_k, -b'_k \in H$ and $a'_{k-1} \geq b'_{k-1} \geq a'_k \geq b'_k$ for all $k < m$.

A convenient characterization of skeletal algebras is given by the following.

2.5.30 Theorem

(1) *Let $(B, \mathbf{C})$ be a closure algebra and $(B(H), \mathbf{C}_H)$ its skeletal subalgebra. Then*

$$B(H) = \{a \in B : (\exists m \in \omega)(\rho^m a = 0)\}.$$

(2) *A closure algebra $(B, \mathbf{C})$ is a skeletal algebra if and only if each element $a \in B$ has finite rank.*

Proof For part (1), if $a \in B(H)$, then for some $m \in \omega$ we have $a = \bigvee\limits_{k<m} (a_k - b_k)$ where $-a_k, -b_k \in H$ and so a_k and b_k are closed. Then by Proposition 2.5.27, $\rho^m a = 0$. Conversely, if $\rho^m a = 0$, then a is of rank m, so by Corollary 2.5.21, $a = \bigvee_{k<m} \big(\mathbf{C}\rho^k a - \mathbf{C}(\mathbf{C}\rho^k a - \rho^k a)\big)$. Since $-\mathbf{C}\rho^k a, -\mathbf{C}(\mathbf{C}\rho^k a - \rho^k a) \in H$, we conclude that $a \in B(H)$.

Part (2) follows from part (1) by the definition of skeletal algebras. □

Bibliographic notes. The notion of a relatively complete subalgebra of a Boolean algebra was introduced by P. Halmos in [14]. Construction 2.5.7 is due to M. Stone and G. Birkhoff. Results 2.5.8, 2.5.9, 2.5.11 were obtained by the author [9, 12] and are based on [10] (see also [11]). The ρ-operator (Definition 2.5.12), classification of elements (Definition 2.5.15), and the results related to these notions are due to the author and are published here for the first time.

References

1. Balbes, R., & Dwinger, Ph. (1974). *Distributive lattices*. Columbia: University of Missouri Press.
2. Birkhoff, G. (1967). *Lattice theory* (3rd ed., Vol. XXV). Providence: American Mathematical Society Colloquium Publications, American Mathematical Society.
3. Blok, W. J. (1976). Varieties of interior algebras. Ph.D. thesis, University of Amsterdam.
4. Blok, W. J., & Dwinger, Ph. (1975). Equational classes of closure algebras. I. *Indagationes Mathematicae*, *37*, 189–198.
5. Dowker, C. H., & Papert, D. (1966). Quotient frames and subspaces. *Proceedings of the London Mathematical Society*, *3*(16), 275–296.
6. Dwinger, Ph. (1961). *Introduction to Boolean algebras* (Vol. 40). Hamburger Mathematische Einzelschriften. Würzburg: Physica-Verlag.
7. Esakia, L. L. (1976). On modal "companions" of superintuitionistic logics (Russian). In *VII Soviet Symposium on Logic (Kiev, 1976)* (pp. 135–136).
8. Esakia, L. L. (1976). On the Dummett-Lemmon conjecture (Russian). In *IV Soviet Conference In Mathematical Logic (Chisinau, 1976)* (p. 160).
9. Esakia, L. L. (1979). On the theory of modal and superintuitionistic systems (Russian). *Logical inference* (pp. 147–172). Moscow: Nauka.
10. Esakia, L. L. (1974). Topological Kripke models. *Soviet Mathematics Doklady*, *15*, 147–151.
11. Esakia, L. L. (1981). Diagonal constructions, Löb's formula, and Cantor's scattered spaces (Russian). In *Studies in Logic and Semantics, Metsniereba, Tbilisi* (pp. 128–143).
12. Esakia, L. L. (1984). On the variety of Grzegorczyk algebras. *Sel. Sov. Math.*, *3*, 343–366.
13. Esakia, L. L., & Meskhi, S. (1977). Five critical modal systems. *Theoria*, *43*, 52–60.
14. Halmos, P. R. (1962). *Algebraic logic*. New York: Chelsea Publishing Co.
15. Kuznetsov, A. V. (1971). On superintuitionistic logics (Russian). *Mathematical Investigations*, *6*, 75–122.
16. Maksimova, L. L., & Rybakov, V. V. (1974). The lattice of normal modal logics. *Algebra and Logic*, *13*, 105–122.
17. McKinsey, J. C. C., & Tarski, A. (1944). The algebra of topology. *Annals of Mathematics*, *2*(45), 141–191.
18. McKinsey, J. C. C., & Tarski, A. (1948). Some theorems about the sentential calculi of Lewis and Heyting. *Journal of Symbolic Logic*, *13*, 1–15.
19. Rasiowa, H., & Sikorski, R. (1963). *The mathematics of metamathematics* (Vol. 41). Monografie Matematyczne. Warsaw: Państwowe Wydawnictwo Naukowe.
20. Sikorski, R. (1960). *Boolean algebras* (Vol. 25). Ergebnisse der Mathematik und ihrer Grenzgebiete, N. F. Berlin: Springer.
21. Thomason, S. K. (1972). Semantic analysis of tense logics. *Journal of Symbolic Logic*, *37*, 150–158.

Chapter 3
Duality Theory: Hybrids

3.1 The Hybrid of Topology (Stone) and Order (Kripke)

Let X be a Stone space. Clearly each binary relation R on X for which the sets $R(x) = \{y \in X : xRy\}$, for each $x \in X$, are nonempty and closed gives rise to (or can be thought of as) a map $\rho : X \to \exp X$ from the space X to the space $\exp X$ of nonempty closed subsets of X equipped with the exponential topology.[1] It is natural to regard R as **continuous** if the corresponding map $\rho : X \to \exp X$ is continuous.

Let the set X be equipped with a quasi-order R and a topology Ω making X a compact Hausdorff zero-dimensional space. The resulting triple (X, Ω, R) is therefore a "mixture" of a Kripke model (X, R), i.e., a set of "worlds" X equipped with a relation of "reachability" R, and a Stone space (X, Ω). We adopt the following fundamental definition.

3.1.1 Definition A triple (X, Ω, R) is said to be a **hybrid** if:

(1) (X, Ω) is a Stone space;
(2) (X, R) is a quasi-order;
(3) the relation R is continuous (the condition "connecting" topology and order).

We call (X, Ω) the **topological component** and (X, R) the **order component** of the hybrid (X, Ω, R).

The following theorem characterizes hybrids from various points of view.

3.1.2 Theorem *Let (X, Ω) be a Stone space and (X, R) a quasi-ordered set. The following conditions are equivalent:*

(I) *(X, Ω, R) is a hybrid;*
(II) (a) *$R(x)$ is a closed set for each $x \in X$;*
(b) *for each closed (resp. open) set $A \subseteq X$, the set $R^{-1}(A) = \{x \in X : R(x) \cap A \neq \varnothing\}$ is closed (resp. open);*

[1] *Editorial note*: For the definition of the exponential (Vietoris) topology, see Definition 1.3.4.

G. Bezhanishvili and W. H. Holliday (eds.), *Heyting Algebras*,
Trends in Logic 50, https://doi.org/10.1007/978-3-030-12096-2_3

(III) (a) *$R(x)$ is a closed set for each $x \in X$;*
(b) *if A is clopen in the space X, then so is $R^{-1}(A)$.*

(IV) ***Separation condition***

(a) *If $y \notin R(x)$, then there exists a cone partition of the space X, i.e., a partition of X into two disjoint open sets U_1 and U_2 such that $x \in U_1$ and U_1 is an upper cone, and $y \in U_2$ and U_2 is a lower cone.*
(b) *For any $x \in X$ and any open $V \subseteq X$ such that $R(x) \cap V \neq \varnothing$, there is an open neighbourhood U of x such that $R(z) \cap V \neq \varnothing$ for each $z \in U$.*

(V) ***Closed graph condition***

(a) *The graph of the relation R is closed in the topological space $X \times X = X^2$;*
(b) *the set $R^{-1}(U)$ is open for each open set U.*

(VI) ***Cones and topological closure***

(a) *$R^{-1}(x)$ is closed for each $x \in X$;*
(b_1) *the smallest closed set containing a cone is a cone;*
(b_2) *the smallest cone containing a closed set is closed.*

(VII) ***Condition of commutation of operators R and $\mathbf{C}$***

(a) *$R^{-1}(x)$ is closed for each $x \in X$;*
(b) *$R\mathbf{C}A = \mathbf{C}RA$ for each set $A \subseteq X$, where $\mathbf{C}A$ is the topological closure of A in the space (X, Ω).*[2]

The theorem is based on a number of lemmas. For brevity and without the danger of confusion, we will often write X instead of (X, Ω) and (X, R) instead of (X, Ω, R).

3.1.3 Lemma *(X, Ω, R) is a hybrid if and only if:*

(1) *for each open set U, the set $R^{-1}(U)$ is open;*
(2) *for each closed set F, the set $R^{-1}(F)$ is closed;*
(3) *$R(x)$ is closed for each $x \in X$.*

Proof Clearly $\rho : X \to \exp X$ is well defined if and only if $R(x)$ is closed for each $x \in X$. Recall that a map f from a topological space Y to a topological space Z is continuous if and only if the f-preimage of each set from an open subbasis of Z is open. A subbasis of the space $\exp X$ consists of the sets of the form $\mathbb{F}_1(U) = \{F \in \exp X : F \cap U \neq \varnothing\}$ and $\mathbb{F}_2(U) = \{F \in \exp X : F \subseteq U\}$, where U is an open subset of X. It is easy to check that:

[2] *Editorial note*: Following the policy for omission of parentheses explained in the Preface, in contexts involving $\mathbf{C}$, parentheses will sometimes be omitted after R and R^{-1} for readability.

$$\text{(a)} \quad \begin{aligned} \rho^{-1}(\mathbb{F}_1(U)) &= \{x \in X : \rho(x) \in \mathbb{F}_1(U)\} \\ &= \{x \in X : \rho(x) \cap U \neq \varnothing\} \\ &= \{x \in X : R(x) \cap U \neq \varnothing\} \\ &= R^{-1}(U); \end{aligned}$$

$$\text{(b)} \quad \begin{aligned} -\rho^{-1}(\mathbb{F}_2(U)) &= -\{x \in X : \rho(x) \subseteq U\} \\ &= -\{x \in X : \rho(x) \cap -U = \varnothing\} \\ &= \{x \in X : R(x) \cap -U \neq \varnothing\} \\ &= R^{-1}(-U). \end{aligned}$$

Let ρ be continuous. Then $\rho^{-1}(\mathbb{F}_1(U))$ is open for each open set $U \subseteq X$. By (a), $\rho^{-1}(\mathbb{F}_1(U)) = R^{-1}(U)$. Therefore, $R^{-1}(U)$ is open if U is open, i.e., (1) holds. Also, $\rho^{-1}(\mathbb{F}_2(U))$ is open, i.e., $-\rho^{-1}(\mathbb{F}_2(U))$ is closed. By (b), $-\rho^{-1}(\mathbb{F}_2(U)) = R^{-1}(-U)$, so the R-preimage of a closed set is closed, i.e., (2) holds. Conversely, suppose conditions (1)–(2) of the lemma hold. Then $R^{-1}(U)$ is open for each open $U \subseteq X$. By (a), $R^{-1}(U) = \rho^{-1}(\mathbb{F}_1(U))$. Moreover, the set $-U$ is closed. By condition (2), $R^{-1}(-U)$ is closed. By (b), $R^{-1}(-U) = -\rho^{-1}(\mathbb{F}_2(U))$. Therefore, $\rho^{-1}(\mathbb{F}_2(U))$ is open. The continuity of ρ has thus been established. □

3.1.4 Lemma *Let X be a Stone space and R a quasi-order on X such that:*

(a) *for each $x \in X$, the set $R(x)$ is closed;*
(b) *for each clopen set A, the set $R^{-1}(A)$ is clopen.*

If $y \notin R(x)$, then there exists a partition of the space X into two open sets U_1 and U_2 such that $x \in U_1$ and U_1 is an upper cone, and $y \in U_2$ and U_2 is a lower cone.

Proof Let $x, y \in X$ and $y \notin R(x)$. Then $y \in -R(x)$, and $-R(x)$ is an open lower cone since $R(x)$ is a closed upper cone. Because the space is zero-dimensional, there exists a clopen set A such that $y \in A \subseteq -R(x)$. Let $U_2 = R^{-1}(A)$ and $U_1 = -R^{-1}(A)$. Clearly $y \in U_2$ and U_2 is a lower cone. By condition (b) of the lemma, $R^{-1}(A)$ is clopen. It is easy to see that $x \notin R^{-1}(A)$; for otherwise, $R(x) \cap A \neq \varnothing$, which is impossible since $A \subseteq -R(x)$. Thus, the set $U_1 = -R^{-1}(A)$ is a clopen upper cone such that $x \in U_1$. □

3.1.5 Lemma *Let X be a Stone space and R a quasi-order on X such that if $y \notin R(x)$, then there exists a partition of the space X into two open sets U_1 and U_2 such that $x \in U_1$ and U_1 is an upper cone, and $y \in U_2$ and U_2 is a lower cone. Then the set $G = \{(x, y) : y \in R(x)\}$ is closed in the product $X \times X = X^2$.*

Proof If G is not closed in X^2, then there is $(x, y) \in X^2$ such that

(1) $(x, y) \in \mathbf{C}G$ and $(x, y) \notin G$ (i.e., $y \notin R(x)$).

Let $\{U_1, U_2\}$ be the partition of X into two open sets such that $x \in U_1$ and U_1 is an upper cone, and $y \in U_2$ and U_2 is a lower cone. Since $U_1 \times U_2$ is an open neighborhood of (x, y), and (x, y) belongs to the closure of G (in X^2), we have $(U_1 \times U_2) \cap G \neq \varnothing$, i.e., there are $x' \in U_1$ and $y' \in U_2$ such that $(x', y') \in G$, so $x'Ry'$. As U_1 is an upper cone, from $x' \in U_1$ and $x'Ry'$ it follows that $y' \in U_1$. Therefore, $y' \in U_1$ and $y' \in U_2$, i.e., $U_1 \cap U_2 \neq \varnothing$. The obtained contradiction concludes the proof. □

3.1.6 Lemma *Let X be a Stone space and R a quasi-order on X such that the graph of G is closed in X^2. Then:*

(a) *the smallest cone containing a closed set is closed;*
(b) *$R^{-1}(x)$ is closed in X for each $x \in X$.*

Proof First we show that if a set F is closed, then $R(F)$ is closed. Since G is closed in X^2, G is a Stone space with respect to the subspace topology. The projection $p : G \to X$ (where $p(x, y) = y$ for $x \in G$) is a continuous map from the Stone space G onto the Stone space X and hence is a closed map. Let F be a closed subset of X. Then $(F \times X) \cap G$ is closed in G, and as p is closed, the set $p((F \times X) \cap G) = \{y \in X : (\exists\,(x_0, y_0) \in (F \times X) \cap G)(p(x_0, y_0) = y)\}$ is closed in X. It is easy to see that the following equalities hold:

$$\begin{aligned}
&\Big\{y \in X : (\exists\,(x_0, y_0) \in (F \times X) \cap G)\,\big(p(x_0, y_0) = y\big)\Big\}\\
&\quad = \Big\{y \in X : (\exists\, x_0 \in F)\,\big((x_0, y) \in (F \times X) \cap G\big)\Big\}\\
&\qquad = \Big\{y \in X : (\exists\, x_0 \in F)\,\big(x_0 \in R^{-1}(y)\big)\Big\} = R(F).
\end{aligned}$$

Therefore, $R(F)$ is closed. Thus, if F is a closed set, then the smallest cone containing F (i.e., the set $R(F)$) is closed.

A similar argument shows that if F is a closed set, then $R^{-1}(F)$ is closed. In particular, as each point of X is closed, so is $R^{-1}(x)$. □

3.1.7 Lemma *Let X be a Stone space and R a quasi-order on X such that:*

(1) *the smallest closed set containing a cone is a cone;*
(2) *the smallest cone containing a closed set is closed.*

Then $R\mathbf{C}A = \mathbf{C}RA$ for each subset A of X, where $\mathbf{C}$ is the operation of topological closure in X.

Proof It is easy to see that condition (1) can be written as the following inclusion:

(1_0) $R\mathbf{C}RA \subseteq \mathbf{C}RA$.

Since R is reflexive, $A \subseteq R(A)$, so $\mathbf{C}A \subseteq \mathbf{C}RA$, and hence $R\mathbf{C}A \subseteq R\mathbf{C}RA$. This together with (1_0) and the transitivity of $\subseteq$ gives $R\mathbf{C}A \subseteq \mathbf{C}RA$. Condition (2) can be rewritten as the following inclusion:

(2_0) $\mathbf{C}R\mathbf{C}A \subseteq R\mathbf{C}A$.

Since $A \subseteq \mathbf{C}A$, we have $RA \subseteq R\mathbf{C}A$, so $\mathbf{C}RA \subseteq \mathbf{C}R\mathbf{C}A$. By ($2_0$) and the transitivity of $\subseteq$, we obtain $\mathbf{C}RA \subseteq R\mathbf{C}A$. The two inclusions $R\mathbf{C}A \subseteq \mathbf{C}RA$ and $\mathbf{C}RA \subseteq R\mathbf{C}A$ give the required equality. □

3.1.8 Lemma *Let X be a Stone space and R a quasi-order on X such that:*

(1) *$R^{-1}(x)$ is closed for each $x \in X$;*
(2) *$R\mathbf{C}A = \mathbf{C}RA$ for each subset A of X.*

Then:

(a) *$R(x)$ is closed for each $x \in X$;*
(b) *for each closed (resp. open) set A in X, the set $R^{-1}(A)$ is closed (resp. open).*

Proof First we show that if F is closed, then $R(F)$ is closed. Since F is closed, $\mathbf{C}F = F$. By (2), $\mathbf{C}RF = R\mathbf{C}F = R(F)$. Therefore, $R(F)$ is closed. As each singleton $\{x\}$ is closed, it follows that $R(x)$ is closed.

Next we show that if U is open, then $R^{-1}(U)$ is open. Since $R^{-1}(U)$ is a lower cone, $-R^{-1}(U)$ is an upper cone. Moreover, $U \cap -R^{-1}(U) = \varnothing$. As U is open, $U \cap \mathbf{C}{-R^{-1}(U)} = \varnothing$. By (2), the closure of each upper cone is an upper cone, so $\mathbf{C}{-R^{-1}(U)}$ is an upper cone. Therefore, $R^{-1}(U) \cap \mathbf{C}{-R^{-1}(U)} = \varnothing$, and so $\mathbf{C}{-R^{-1}(U)} \subseteq -R^{-1}(U)$. This implies that $-R^{-1}(U)$ is closed, and hence $R^{-1}(U)$ is open.

It is left to show that if F is closed, then $R^{-1}(F)$ is closed. For this it is sufficient to show that $R^{-1}(F) = \bigcup_{x \in F} R^{-1}(x)$ is compact. Let $\mathbb{F} = \{U_j : j \in J\}$ be an open cover of $R^{-1}(F)$. Clearly $\mathbb{F}$ is an open cover of $R^{-1}(x)$ for each $x \in F$. By (1), $R^{-1}(x)$ is closed and hence compact. Therefore, there is a finite subcover, say $\{U_j : j \in J_x\}$. Let $U_x = \bigcup\{U_j : j \in J_x\}$ and $U_x^0 = -R{-U_x}$. As we already saw, the R-image of a closed set is closed. This implies that U_x^0 is open. From $R^{-1}(x) \subseteq U_x$ it follows that $x \in U_x^0$. Consequently, $\{U_x^0 : x \in F\}$ is an open cover of the closed set F, so there is a finite subcover, say $\{U_{x_1}^0, \ldots, U_{x_k}^0\}$. But then $\{U_j : j \in J_{x_1} \cup \cdots \cup J_{x_k}\}$ is a finite subcover of $R^{-1}(F)$. Thus, $R^{-1}(F)$ is compact and hence closed. □

Proof of Theorem 3.1.2. By Lemma 3.1.3, conditions (I) and (II) are equivalent.

(II $\Rightarrow$ III) Let A be a clopen subset of X. By (II), the set $R^{-1}(A)$ is closed since A is closed, and $R^{-1}(A)$ is open since A is open.

(III $\Rightarrow$ IV) By Lemma 3.1.4, (III) implies (IVa). We now show (IVb). Since $R^{-1}\big(\bigcup_{i \in I} A_i\big) = \bigcup_{i \in I} R^{-1}(A_i)$, it follows from (IIIb) and the zero-dimensionality of X that the R-preimage of each open set is open. Let $x \in X$ and V be an open set such that $R(x) \cap V \neq \varnothing$. Let $U = R^{-1}(V)$. Since V is open, U is open. As $R(x) \cap V \neq \varnothing$, we have $x \in U$. If $z \in U$, then there exists $y \in V$ such that $z \in R^{-1}(y)$. Therefore, $y \in R(z)$, and so $R(z) \cap V \neq \varnothing$.

(IV $\Rightarrow$ V) By Lemma 3.1.5, (Va) follows from (IVa). We now prove (Vb). Let U be an open set. Consider the set $R^{-1}(U)$. Let $z \in R^{-1}(U)$. By (IVb), there exists an open set V such that $z \in V$ and $(\forall\, y \in V)(R(y) \cap U \neq \varnothing)$. The latter condition can be written as $V \subseteq R^{-1}(U)$. Therefore, each point $z \in R^{-1}(U)$ is an interior point, and hence $R^{-1}(U)$ is open.

(V $\Rightarrow$ VI) By Lemma 3.1.6, (VIa) and (VIb$_2$) follow from (Va). It remains to prove (VIb$_1$). For this it is sufficient to show that $\mathbf{C}RA$ is a cone for each $A \subseteq X$. Let $x \in \mathbf{C}RA$ and xRy. If $y \notin \mathbf{C}RA$, then there exists an open set U such that $y \in U$ and $U \cap R(A) = \varnothing$. Since $R(A)$ is a cone, this implies $R^{-1}(U) \cap R(A) = \varnothing$. By (Vb), $R^{-1}(U)$ is open. Therefore, $R^{-1}(U) \cap \mathbf{C}RA = \varnothing$. As xRy, we have $x \in R^{-1}(U)$. Therefore, $R^{-1}(U)$ is an open neighborhood of x entirely missing $\mathbf{C}RA$. Thus, $x \notin \mathbf{C}RA$, a contradiction. Consequently, $\mathbf{C}RA$ is a cone, and so the smallest closed set containing a cone is a cone.

(VI $\Rightarrow$ VII) Follows from Lemma 3.1.7.

(VII $\Rightarrow$ II) Follows from Lemma 3.1.8.

The theorem is proved. □

We give a few "degenerate" examples of hybrids.

(A) Let (X, Ω) be a Stone space and R the identity relation on the set X, i.e, $x \in R^{-1}(y)$ if and only if $x = y$. Then (X, Ω, R) is a hybrid. Indeed, since $R^{-1}(x) = \{x\}$ for each $x \in X$ and all singletons $\{x\}$ are closed (as X is Hausdorff), condition (VIIa) of Theorem 3.1.2 holds. Since for each set A in X, we have $R(A) = A$, the commutation condition (VIIb) obviously holds as well.

(B) Let (X, Ω) be a Stone space and R the relation on X such that $x \in R(y)$ and $y \in R(x)$ for each $x, y \in X$ (i.e., (X, R) is a cluster). Then (X, Ω, R) is a hybrid. Indeed, in this case, $R(A) = X$ for each nonempty subset A of X, and $R(\varnothing) = \varnothing$. Clearly condition (VI) of Theorem 3.1.2 holds.

(C) Let (X, R) be a finite quasi-ordered set. Equip X with the discrete topology. Then it is obvious that (X, Ω, R) is a hybrid.

We conclude this section by fixing some notation and terminology.

3.1.9

(1) The class (and the category) of all hybrids will be denoted by HYB.[3]
(2) HYB^+ is the class (and the category) of all **strict** (or partially ordered) hybrids, i.e., hybrids (X, Ω, R) such that R is a partial order.
(3) ST is the class (and the category) of Stone spaces.
(4) QO (resp. QOF) is the category of all (resp. all finite) quasi-ordered sets and strictly isotone maps.
(5) PO (resp. POF) is the category of all (resp. all finite) partially ordered sets and strictly isotone maps.

Bibliographic notes. Hybrids (Definition 3.1.1) were introduced originally by the author in [4] under the name of 'perfect Kripke models'. The notion was motivated by the papers of Jónsson and Tarski [10] and Halmos [13].

---

[3] *Editorial note*: The morphisms of the category HYB will be introduced in Definition 3.3.1.

3.2 Fundamental Properties of Hybrids

3.2.1 Theorem *Let* (X, R) *be a strict hybrid and* F *a closed subset of* X. *Then for each* $x \in F$, *there exists a maximal (resp. minimal)* $y \in F$ *such that* xRy *(resp.* yRx*).*

Proof Suppose that there exists $x \in F$ such that $\max(F \cap R(x)) = \varnothing$. Let Y be a maximal chain contained in $R(x) \cap F$.[4] Consider the family $\mathbb{F} = \{R(y) \cap F : y \in Y\}$ of nonempty closed sets. Clearly $\mathbb{F}$ is a chain with respect to inclusion and by compactness of X, we have $\bigcap\{R(y) \cap F : y \in Y\} \neq \varnothing$. Let $z \in \bigcap\{R(y) \cap F : y \in Y\}$, i.e., for each $y \in Y$, we have yRz and $z \in F$. Then $z \in \max Y$ because Y is a maximal chain. Therefore, $z \in \max(R(x) \cap F)$, a contradiction. The dual part of the theorem is proved similarly. □

3.2.2 Corollary *Let* $(X, R) \in \mathsf{HYB}^+$ *and* F *be a nonempty closed subset of* X. *Then* $\max F \neq \varnothing$ *and* $\min F \neq \varnothing$.

Remark to Theorem 3.2.1 **and Corollary** 3.2.2. It is easy to check that Theorem 3.2.1 and Corollary 3.2.2 hold for arbitrary hybrids provided in the statement the prefix 'quasi-' is added to the words 'maximal' and 'minimal'. It is appropriate here to recall Definition 1.4.7 and especially the informal remarks after it.

3.2.3 Theorem *Let* $(X, R) \in \mathsf{HYB}^+$. *Then the set* $\max X$ *is closed in* X.

Proof Let $y \in X$ and $y \notin \max X$. Consider $x \in \max X$ such that yRx (see Theorem 3.2.1). Then $y \notin R(x)$. By Theorem 3.1.2, there is a clopen upper set A such that $x \in A$ and $y \notin A$. Clearly $y \in R^{-1}(A) - A$ and $x \notin R^{-1}(A) - A$. Since $R^{-1}(A)$ is clopen, so is $R^{-1}(A) - A$. Obviously $(R^{-1}(A) - A) \cap \max A = \varnothing$. Therefore, y has an open neighborhood $R^{-1}(A) - A$ disjoint from $\max X$. Thus, $y \notin \mathbf{C} \max X$, where $\mathbf{C} \max X$ is the topological closure of $\max X$. □

However, the dual statement of Theorem 3.2.3 is false as the following theorem shows.

3.2.4 Theorem *There exists a strict hybrid* (X, Ω, R) *such that the set* $\min X$ *is not closed.*

Proof Let X be the one-point compactification of an infinite discrete space X_0, and let $z \in X - X_0$, i.e., z is the only compactification point. It is known that X is a Stone space.[5] Define a partial order R on X as follows:

(1) xRx for each $x \in X$ (reflexivity of R);
(2) $R^{-1}(x) = \{x\}$ for each $x \in X_0$;
(3) $R^{-1}(z) = X$.

---

[4] *Editorial note*: The existence of such a maximal chain follows from Proposition 1.4.10.

[5] *Editorial note*: See, e.g., Example D on p. 26 of [18].

Therefore, z is the greatest point of X, and $\min X = X_0$. It is easy to check that (X, Ω, R) is a strict hybrid and that the set $\min X = X_0$ is not closed in the space X (recall that z is the only non-isolated point of the space). □

Recall that a set A in a partially ordered set (X, R) is said to be an **antichain** if any two distinct points x, y of A are incomparable, i.e., $x \notin R(y)$ and $y \notin R(x)$.

3.2.5 Corollary *There exist a strict hybrid (X, R), an open set $U \subseteq X$, and an antichain $A \subseteq X$ such that:*

(1) *the set $R(U)$ is not open in the space X;*
(2) *the closure of the antichain A (i.e., the set $\mathbf{C}A$) in X is not an antichain.*

Proof Let (X, R) be the strict hybrid constructed in the proof of Theorem 3.2.4.

(1) Set $U = \{x\}$, where $x \in X_0$. Clearly U is open, and $R(U) = R(x) = \{x, z\}$, which is not open.

(2) Take the set X_0 as our antichain A. The closure $\mathbf{C}X_0$ of X_0 is equal to X, but X is not an antichain. □

For a hybrid (X, Ω, R) and a subset Y of X, let (Y, Ω_Y, R_Y) denote the set Y equipped with the subspace topology Ω_Y and the order R_Y which is the restriction of R to Y, i.e., $R_Y(x) = R(x) \cap Y$ for each $y \in Y$.

3.2.6 Theorem *Let $(X, \Omega, R) \in \mathsf{HYB}^+$ and Y be a clopen set in (X, Ω). Then (Y, Ω_Y, R_Y) is a strict hybrid.*

Proof It is well known that (Y, Ω_Y) is a Stone space. Moreover, if $x \in Y$, then the set $R_Y(x) = R(x) \cap Y$ is closed in the space Y. Now let A be a clopen set in (Y, Ω_Y). Then $A = A_0 \cap Y$ for some clopen A_0 in (X, Ω). Since (X, Ω, R) is a hybrid, by Theorem 3.1.2, $R^{-1}(A_0 \cap Y)$ is clopen in the space (X, Ω). The remainder of the proof follows from the following easily verifiable equalities:

$$\begin{aligned} R_Y^{-1}(A_0 \cap Y) &= \bigcup \{R_Y^{-1}(x) : x \in A_0 \cap Y\} \\ &= \bigcup \{R^{-1}(x) \cap Y : x \in A_0 \cap Y\} \\ &= Y \cap \bigcup \{R^{-1}(x) : x \in A_0 \cap Y\} \\ &= Y \cap R^{-1}(A_0 \cap Y). \end{aligned}$$

□

3.2.7 Corollary *Let Y be a clopen subset of a strict hybrid (X, Ω, R). Then the set* $\max Y$ *is closed in (X, Ω).*

Proof By Theorem 3.2.6, (Y, Ω_Y, R_Y) is a hybrid. Therefore, by Theorem 3.2.3, the set $\max Y$ is closed in (Y, Ω_Y). Since Y is closed in (X, Ω), $\max Y$ is closed in (X, Ω). □

If (X, R) is a partially ordered set, then we call a set of the form $R(x) \cup R^{-1}(x)$, for $x \in X$, a **bow** (with the **knot** x).

3.2.8 Theorem *Let Y be a chain of a strict hybrid (X, R). Then its topological closure $\mathbf{C}Y$ is also a chain.*

Proof Let Y be a chain in (X, R) and $y \in \mathbf{C}Y$. To show that y is comparable with each $x \in Y$, i.e., that $y \in R(x) \cup R^{-1}(x)$ for each $x \in Y$, suppose $y \notin R(x) \cup R^{-1}(x)$ for some $x \in Y$. The bow $R(x) \cup R^{-1}(x)$ is closed and $Y \subseteq R(x) \cup R^{-1}(x)$. Therefore, $y \in -(R(x) \cup R^{-1}(x))$, $-(R(x) \cup R^{-1}(x))$ is open, and $Y \cap -(R(x) \cup R^{-1}(x)) = \varnothing$. Thus, $y \notin \mathbf{C}Y$, a contradiction. Consequently, y is comparable with each point of Y.

Let $x, y \in \mathbf{C}Y$ and $x, y \notin Y$. We will show that x and y are comparable. If not, then $x \notin R(y) \cup R^{-1}(y)$, i.e., $x \in -(R(y) \cup R^{-1}(y))$. Note that $Y \subseteq R(y) \cup R^{-1}(y)$. Therefore, $x \notin \mathbf{C}Y$, a contradiction. Thus, $x \in R(y) \cup R^{-1}(y)$. □

3.2.9 Corollary *Let $(X, R) \in \mathsf{HYB}^+$ and Y be a maximal chain in (X, R). Then Y is closed in X.*

Proof Since Y is a chain, $\mathbf{C}Y$ is also a chain by Theorem 3.2.8. Then since $Y \subseteq \mathbf{C}Y$ and Y is a maximal chain, we have $Y = \mathbf{C}Y$, so Y is closed. □

3.2.10 Theorem *Let (X, Ω, R) be a strict hybrid and Y a closed chain. Then (Y, Ω_Y, R_Y) is a linear strict hybrid (i.e., a strict hybrid whose order component (Y, R_Y) is a chain).*

Proof Since Y is closed in (X, Ω), (Y, Ω_Y) is a Stone space. As $R_Y(x) = Y \cap R(x)$ for $x \in Y$, the set $R_Y(x)$ is closed in (Y, Ω_Y). Let A be clopen in (X, Ω). Then $A \cap Y$ is clopen in (Y, Ω_Y). Consider the set $A_0 = R_Y^{-1}(A \cap Y) = Y \cap R^{-1}(A \cap Y)$. Clearly A_0 is closed in (Y, Ω_Y). We will prove that A_0 is open in (Y, Ω_Y). Let $x \in \max(A \cap Y)$. Since (Y, R_Y) is a chain, we have $R_Y^{-1}(A \cap Y) = R_Y^{-1}(x)$. The set $Y_0 = (R(x) - \{x\}) \cap Y$ is contained in the closed set $R(x) \cap Y$ and is open in (Y, Ω_Y) since $Y_0 = Y - R_Y^{-1}(x)$. Consider the set $\mathbf{C}_Y Y_0$, i.e., the topological closure of Y_0 in (Y, Ω_Y). Since $Y_0 \subseteq R(x) \cap Y = R_Y(x)$ and the set $R(x) \cap Y$ is closed in (Y, Ω_Y), we have $\mathbf{C}_Y Y_0 \subseteq R(x) \cap Y$. Moreover, $x \notin \mathbf{C}_Y Y_0$ since $A \cap Y$ is clopen in Y, $x \in A \cap Y$, and $(A \cap Y) \cap Y_0 = \varnothing$. Let $x_0 \in \min \mathbf{C}_Y Y_0$. Then x is covered by x_0 (i.e., $x R_Y x_0$, and if $x R_Y z$ and $z R_Y x_0$, then $x = z$ or $z = x_0$). The set $R_Y(x_0)$ is closed since $R_Y(x_0) = Y \cap R(x_0)$, and so the set $Y - R_Y(x_0) = R_Y^{-1}(x) = R_Y^{-1}(A \cap Y)$ is open in (Y, Ω_Y). The theorem is proved. □

If (X, R) is a partially ordered set and $Y \subseteq X$, let $\sup Y$ (resp. $\inf Y$) denote the least (resp. greatest) element of the set $\{x \in X : (\forall y \in Y)(y R x)\}$ (resp. of the set $\{x \in X : (\forall y \in Y)(x R y)\}$).

3.2.11 Lemma *Let Y be a nonempty chain in a strict hybrid (X, R). Then* $\sup Y$ *and* $\inf Y$ *exist and* $\sup Y, \inf Y \in \mathbf{C}Y$*, where $\mathbf{C}Y$ is the closure of Y in X.*

Proof Let Y be a nonempty chain in (X, R). If $\max Y \neq \varnothing$, then clearly $\max Y = \{x\}$ and $x = \sup Y$. It is also clear that $x \in Y$, and since $Y \subseteq \mathbf{C}Y$, we have $x \in \mathbf{C}Y$. Let $\max Y = \varnothing$. By Theorem 3.2.8, $\mathbf{C}Y$ is a closed chain in the hybrid (X, R). Therefore, by Theorem 3.2.1, $\max(\mathbf{C}Y) \neq \varnothing$. Let $x \in \max \mathbf{C}Y$. Then $x = \sup Y$. Indeed, since $Y \subseteq \mathbf{C}Y$ and $x \in \max \mathbf{C}Y$, for each $y \in Y$, yRx. Let z be such that yRz for all $y \in Y$. We will show that xRz, i.e., that $z \in R(x)$. Suppose that $z \notin R(x)$. Then the chain $Y_0 = R^{-1}(z) \cap \mathbf{C}Y$ is a closed set, $x \notin Y_0$, and $Y \subseteq Y_0$. Since $Y \subseteq Y_0$, $\mathbf{C}Y \subseteq \mathbf{C}Y_0 = Y_0$. As $x \notin Y_0$, we also have $x \notin \mathbf{C}Y$, which together with $x \in \mathbf{C}Y$ yields a contradiction. Thus, $z \in R(x)$, and so $x = \sup Y$. The existence of $\inf Y$ is proved similarly. □

If (X, R) is a partially ordered set and Y is a chain in (X, R), let $\widetilde{Y}$ denote the **completion** of the chain Y, i.e., for each $x \in X$, $x \in \widetilde{Y}$ if and only if there exists a subset Y_0 of Y such that $\sup Y_0 = x$ or $\inf Y_0 = x$.

3.2.12 Theorem *Let (X, R) be a strict hybrid and Y a chain in (X, R). Then the completion $\widetilde{Y}$ of Y coincides with its topological closure $\mathbf{C}Y$.*

Proof We show that $\widetilde{Y} = \mathbf{C}Y$. If the chain Y is finite, then obviously $\widetilde{Y} = Y = \mathbf{C}Y$. Let Y be infinite, $Y_0 \subseteq Y$, and $\sup Y_0 = x$. By Lemma 3.2.11, $x \in \mathbf{C}Y_0$. Since $\mathbf{C}Y_0 \subseteq \mathbf{C}Y$, we have $x \in \mathbf{C}Y$. Similarly, if $x = \inf Y_0$, then $x \in \mathbf{C}Y$. Therefore, $\widetilde{Y} \subseteq \mathbf{C}Y$. To show that $\mathbf{C}Y \subseteq \widetilde{Y}$, let $x \in \mathbf{C}Y$ and $x \notin Y$. Consider the chains $Y_0 = R^{-1}(x) \cap Y$ and $Y_1 = R(x) \cap Y$. Let $x_0 = \sup Y_0$ and $x_1 = \inf Y_1$. Suppose $x_0 \neq x$ and $x \neq x_1$. Obviously $x_0 \in R^{-1}(x)$ and $x_1 \in R(x)$. Let $Z_0 = R^{-1}(x_0) \cap \mathbf{C}Y$ and $Z_1 = R(x_1) \cap \mathbf{C}Y$. Then Z_0, Z_1 are closed, $Y \subseteq Z := Z_0 \cup Z_1$, and the set Z is closed and contained in $\mathbf{C}Y$. Moreover, $x \notin Z$. Thus, $-Z$ is an open neighborhood of x, and $-Z \cap Y = \varnothing$. Hence, $x \notin \mathbf{C}Y$, a contradiction. So $x = x_0$ or $x = x_1$, and so $\mathbf{C}Y \subseteq \widetilde{Y}$. Consequently, $\widetilde{Y} = \mathbf{C}Y$. □

3.2.13 Definition Let (X, R) be a partially ordered set, $x, y \in X$, $x \neq y$, and xRy. The ordered pair (x, y) is said to be a **gap** in (X, R) if y covers x, i.e., if $z \in R(x) \cap R^{-1}(y)$ implies $z = x$ or $z = y$. A point $x \in X$ is said to be a **gap** in (X, R) if either $x \in \max X$ or there exists $y \in X$ such that (x, y) is a gap in (X, R). Let $\mathrm{Gap}(X, R)$ denote the set of all points of X that are gaps. A partially ordered set (X, R) is said to have **enough gaps** if each "interval" $R(x) \cap R^{-1}(y)$ (where $x \neq y$ and xRy) contains a gap, i.e., if there is a gap (x', y') in (X, R) such that $x', y' \in R(x) \cap R^{-1}(y)$.

3.2.14 Lemma *Let (X, Ω, R) be a linear strict hybrid. Then the linearly ordered set (X, R) has enough gaps.*

Proof Let $x \neq y$ and xRy. We will find a gap in the interval $R(x) \cap R^{-1}(y)$. Since R is anti-symmetric, $x \notin R(y)$. Therefore, there is a clopen lower set A such that $x \in A$ and $y \notin A$. Clearly $-A$ is a clopen upper set, $x \notin -A$, and $y \in -A$. Let $x' \in \max A$ and $y \in \min -A$. Then (x', y') is the gap we sought for. □

3.2.15 Lemma *Let (X, Ω, R) be a linear strict hybrid. Then the family $\mathbb{F}$ of sets of the form $R(x) \cap -R(y)$ and $R(x) \cup -R(y)$, where $x, y \in \mathrm{Gap}(X, R)$, is a reduced field of sets.*

Proof Let $\mathbb{F}_0$ be the smallest field of subsets of X containing all sets of the form $R(x)$, where $x \in \mathrm{Gap}(X, R)$. Clearly $\mathbb{F} \subseteq \mathbb{F}_0$. Since each element $A \in \mathbb{F}_0$ can be represented as $A = \bigcup_{k<m} (A_k - B_k)$, where $m \in \omega$, A_k and B_k are sets of the form $R(x)$ for suitable $x \in \mathrm{Gap}(X, R)$,[6] and $\bigcup_{k<m} (A_k - B_k) = \bigcup_{k<m} (A_k \cap -B_k)$, we have $A \in \mathbb{F}$. Therefore, $\mathbb{F} = \mathbb{F}_0$, and so $\mathbb{F}$ is a field. Moreover, the field $\mathbb{F}$ is reduced. Indeed, let $x, y \in X$ and $x \neq y$. Then since (X, R) is linearly ordered, either xRy or yRx. Without loss of generality we may assume that xRy. By Lemma 3.2.14, there is a gap (x', y') in the interval $R(x) \cap R^{-1}(y)$. We have $R(y') \in \mathbb{F}$, $y \in R(y')$, and $x \notin R(y')$. □

3.2.16 Theorem *Any strict hybrid (X, R) has enough gaps.*

Proof Let $x, y \in X$, $x \neq y$, and xRy. To find a gap in the interval $R(x) \cap R^{-1}(y)$, let Y be a maximal chain in (X, R) such that $x, y \in Y$. By Lemma 3.2.9, Y is closed in X. By Theorem 3.2.10, (Y, R_Y) is a linear strict hybrid. Let (x', y') be a gap in the hybrid (Y, R_Y), where $(x', y') \in Y$ and $x', y' \in R_Y(x) \cap R_Y^{-1}(y)$. Then (x', y') is a gap in the hybrid (X, R). Indeed, let $z \in X$ be such that $z \neq x'$, $z \neq y'$, $x'Rz$, and zRy'. Since the chain Y is maximal, $z \in Y$, contradicting the fact that (x', y') is a gap in (Y, R_Y). □

3.2.17 Theorem *Let (X, Ω, R) be a strict hybrid and Y a closed chain in X. Then the subspace topology of the hybrid (Y, Ω_Y, R_Y) coincides with the classical interval topology of the chain.*

Proof Recall that the "open" intervals of the chain Y are

(1) Y;
(2) $-R_Y(x)$, where $x \in Y$;
(3) $-R_Y^{-1}(x)$, where $x \in Y$;
(4) $-R_Y(x) \cap -R_Y^{-1}(y)$, where yRx and $y \neq x$.

The interval topology Ω_Y^0 of the chain (Y, R_Y) is obtained by taking these open intervals as a basis for the topology. Note that a subset of the chain (Y, R_Y) is open in its interval topology Ω_Y^0 if and only if it is a union of open intervals. Also, recall that in any chain with a least element and a greatest element, the "closed" intervals form a subbasis for closed sets. By Lemma 3.2.11, the chain (Y, R_Y) is a complete lattice, so the space (Y, Ω_Y^0) is compact. Obviously (Y, Ω_Y^0) is Hausdorff. By Lemma 3.2.14, the chain (Y, Ω_Y) has enough gaps. Each closed interval in (Y, R_Y), i.e., each set of the form $R_Y(x) \cap R_Y^{-1}(y)$ where xRy, is closed in (Y, Ω_Y). Therefore, the interval topology Ω_Y^0 is dominated by the topology Ω_Y of the hybrid (Y, Ω_Y, R_Y). But any Hausdorff topology dominated by a compact topology coincides with it.[7] □

---

[6] *Editorial note*: Since (X, R) is a chain, the family $\{R(x) : x \in \mathrm{Gap}(X, R)\}$ is a ring of sets.

[7] *Editorial note*: See, e.g., Corollary 3.1.14 of R. Engelking, *General Topology*, 2nd ed. (Heldermann-Verlag), 1989.

3.2.18 Definition A partially ordered set (X, R) is said to be **Dedekind complete**[8] if for each nonempty upward (resp. downward) directed subset Z of X, $\sup Z$ (resp. $\inf Z$) exists.[9]

3.2.19 Theorem *Each strict hybrid (X, R) is Dedekind complete.*

Proof Let (X, R) be a strict hybrid. By Lemma 3.2.11, each nonempty chain Y of X has $\sup Y$ and $\inf Y$. Now let Z be an arbitrary nonempty upward (resp. downward) directed set. The existence of $\sup Z$ (resp. $\inf Z$) is guaranteed by a theorem of P. Cohn: if all nonempty chains in a partially ordered set have sup and inf, then each nonempty upward (resp. downward) directed set has sup (resp. inf).[10] □

As an immediate consequence of Theorem 3.1.2(IVa), if $x \notin R^{-1}(y)$ in a hybrid (X, R), then there exists a clopen upper cone A such that $x \in A$ and $y \notin A$. The following lemma states that a stronger principle holds.

3.2.20 Lemma *Let (X, R) be a hybrid with $x \in X$ and D a closed lower cone such that $x \notin D$. Then there exists a clopen upper cone A such that $R(x) \subseteq A$ and $A \cap D = \varnothing$.*

Proof Let $x \notin D$. Then $R(x) \subseteq -D$ as $-D$ is an upper cone. For each $y \in D$, we have $x \notin R^{-1}(y)$. Therefore, there is a clopen upper cone A_y such that $x \in A_y$ and $y \notin A_y$. Thus, $\mathbb{F} = \{A_y \cap D : y \in D\}$ is a family of closed subsets of X. Clearly $\bigcap \mathbb{F} = \varnothing$ since for each $y \in D$, we have $y \notin A_y$. By the compactness of X, we have $\bigcap\{A_y \cap D : y \in D_0\} = \varnothing$ for some finite subset D_0 of D. Let $A = \bigcap\{A_y : y \in D_0\}$. Then A is a clopen upper cone, $x \in A$, so $R(x) \subseteq A$, and $A \cap D = \varnothing$. □

3.2.21 Definition

(a) Let (X, Ω) be a Stone space and R a quasi-order on X. We say that the quasi-ordered space (X, Ω, R) satisfies the **Priestley separation axiom** if for any $x, y \in X$ such that $y \notin R(x)$, there exists a partition of X into two open sets X_1 and X_2 such that X_1 is an upper cone and $x \in X_1$, and X_2 is a lower cone and $y \in X_2$.
(b) If in addition the relation R is anti-symmetric (i.e., (X, R) is a partially ordered set), then (X, Ω, R) is said to be a **Priestley space** (or more precisely, a **partially ordered Priestley space**).

3.2.22 Theorem

(1) *Any hybrid (X, Ω, R) satisfies the Priestley separation axiom.*

---

[8] *Editorial note*: This is a nonstandard use of the term 'Dedekind complete', which usually means that each nonempty subset bounded above has a supremum (resp. bounded below has an infimum).

[9] *Editorial note*: A subset Z of X is **upward** (resp. **downard**) **directed** if for all $x, y \in Z$, there is $z \in Z$ such that xRz and yRz (resp. zRx and zRy).

[10] *Editorial note*: See Proposition 5.9 on p. 33 of [3].

(2) *Conversely, if* (X, Ω, R) *is a Priestley space, then* (X, Ω, R) *is a strict hybrid if and only if for any* $x \in X$ *and any open set* V *of* (X, Ω) *such that* $R(x) \cap V \neq \varnothing$, *there exists an open neighborhood* U *of* x *such that* $R(z) \cap V \neq \varnothing$ *for all* $z \in U$.

Proof (1) Follows from Theorem 3.1.2(IVa).

(2) Let (X, Ω, R) be a Priestley space. Then (X, Ω, R) satisfies condition (IVa) of Theorem 3.1.2. Moreover, R is a partial order. Therefore, by Theorem 3.1.2, (X, Ω, R) is a strict hybrid if and only if condition (IVb) of Theorem 3.1.2 holds. □

3.2.23 Definition Let $(X, \Omega, R) \in \mathsf{HYB}$ and $\mathbf{C}$ be the closure operator of the Stone space (X, Ω). Define two operators $\mathbf{C}_1$ and $\mathbf{C}_2$ as follows: $\mathbf{C}_1 A = \mathbf{C}RA$ and $\mathbf{C}_2 A = R^{-1}\mathbf{C}A$ for each subset A of X.

3.2.24 Lemma *Let* (X, Ω, R) *be a hybrid. The operators* $\mathbf{C}_1$ *and* $\mathbf{C}_2$ *satisfy Kuratowski's axioms, where* $A, B \subseteq X$ *and* $i = 1, 2$:

(I) $A \subseteq \mathbf{C}_i A$;
(II) $\mathbf{C}_i\mathbf{C}_i A \subseteq \mathbf{C}_i A$;
(III) $\mathbf{C}_i(A \cup B) = \mathbf{C}_i A \cup \mathbf{C}_i B$;
(IV) $\mathbf{C}_i\varnothing = \varnothing$.

Proof Since $A \subseteq \mathbf{C}A$, $A \subseteq RA$, and $A \subseteq R^{-1}A$ (by reflexivity of R) for any $A \subseteq X$, both operators satisfy axiom (I). The additivity of topological closure and the operators R and R^{-1} implies the additivity of $\mathbf{C}_i$ $(i = 1, 2)$, i.e., axiom (III). Since $\mathbf{C}\varnothing = \varnothing$, $R\varnothing = \varnothing$, and $R^{-1}\varnothing = \varnothing$, axiom (IV) holds. It remains to show that axiom (II) holds.

First we show that $\mathbf{C}_1\mathbf{C}_1 A \subseteq \mathbf{C}_1 A$, i.e., that $\mathbf{C}R\mathbf{C}RA \subseteq \mathbf{C}RA$. By Theorem 3.1.2(VIIb), $\mathbf{C}$ and R commute, so $\mathbf{C}R\mathbf{C}RA = \mathbf{C}\mathbf{C}RRA \subseteq \mathbf{C}RA$ (as R is transitive).

Next we prove that $\mathbf{C}_2\mathbf{C}_2 A \subseteq \mathbf{C}_2 A$, i.e., that $R^{-1}\mathbf{C}R^{-1}\mathbf{C}A \subseteq R^{-1}\mathbf{C}A$. Since $\mathbf{C}A$ is closed, by Theorem 3.1.2(IIb), the set $R^{-1}\mathbf{C}A$ is also closed, i.e., $\mathbf{C}R^{-1}\mathbf{C}A = R^{-1}\mathbf{C}A$. Hence $R^{-1}\mathbf{C}R^{-1}\mathbf{C}A = R^{-1}R^{-1}\mathbf{C}A \subseteq R^{-1}\mathbf{C}A$, where the last inclusion follows from the transitivity of R. □

3.2.25 Lemma *Let* $(X, \Omega, R) \in \mathsf{HYB}$ *and* $\mathbf{C}$ *be the operation of topological closure in* (X, Ω). *Then:*

(1) $\mathbf{C}_1 A = A$ *if and only if* A *is a closed upper cone of the hybrid;*
(2) $\mathbf{C}_2 A = A$ *if and only if* A *is a closed lower cone of the hybrid* (X, Ω, R).

Proof (1) If A is a closed upper cone, then $R(A) = A$ since A is an upper cone and $\mathbf{C}A = A$ since A is closed. Therefore, $\mathbf{C}_1 A = \mathbf{C}RA = A$. Conversely, let $\mathbf{C}_1 A = A$, i.e., $\mathbf{C}RA = A$. Since $\mathbf{C}A \subseteq \mathbf{C}RA$, we have $\mathbf{C}A \subseteq A$. From $RA \subseteq \mathbf{C}RA$ it also follows that $R(A) \subseteq A$. Thus, A is a closed upper cone.

(2) If A is a closed lower cone, then $\mathbf{C}_2 A = R^{-1}\mathbf{C}A = A$. Conversely, let $\mathbf{C}_2 A = A$. We will show that A is a closed lower cone. By Theorem 3.1.2(IIb), if A_0 is

a closed set, then so is $R^{-1}(A_0)$. Therefore, since $\mathbf{C}A$ is closed, so is $R^{-1}\mathbf{C}A$, i.e., $\mathbf{C}R^{-1}\mathbf{C}A = R^{-1}\mathbf{C}A$. By assumption, $R^{-1}\mathbf{C}A = A$. Thus, $\mathbf{C}A = A$, i.e., A is closed. Using $R^{-1}\mathbf{C}A = A$ and $\mathbf{C}A = A$ again yields $R^{-1}(A) = A$, i.e., A is a lower cone. □

3.2.26 Corollary *Each of the families Ω_1 of all open lower cones and Ω_2 of all open upper cones of a hybrid (X, Ω, R) forms a Heyting lattice with respect to inclusion.*

Proof By Lemma 3.2.24, $\mathbf{C}_1$ and $\mathbf{C}_2$ are closure operators. By Lemma 3.2.25, the family Ω_1 (resp. Ω_2) coincides with the family of all open sets of the topology induced by $\mathbf{C}_1$ (resp. $\mathbf{C}_2$). □

We conclude this section by the following remark.

3.2.27 Remark

(1) Theorem 3.2.3 is true for arbitrary hybrids provided in its statement we replace the set max X by the set qmax X of quasi-maximal points.
(2) Theorem 3.2.6 still holds if ‘strict hybrid’ is replaced by ‘hybrid’.
(3) Corollary 3.2.7 holds if ‘strict hybrid’ is replaced by ‘hybrid’ and the set max Y is replaced by qmax Y.
(4) Theorem 3.2.8 is still true if (X, Y) is a hybrid and Y, $\mathbf{C}Y$ are quasi-chains.
(5) Lemma 3.2.9 is also true if (X, R) is a hybrid and Y is a maximal quasi-chain.
(6) Theorem 3.2.10 still holds if (X, Ω, R) is a hybrid, Y is a closed quasi-chain, and (Y, Ω_Y, R_Y) is a quasi-linear hybrid.

Bibliographic notes. There are many publications dedicated to connections between topology and order; we only point out the interesting monograph by Nachbin [15]. In connection with Theorem 3.2.16, it is appropriate to mention the early work of Mostowski and Tarski [14]. For the notion of Dedekind completeness, see [19]. Priestley spaces were introduced (under a different name) in the important paper [16] dedicated to distributive lattices.

3.3 The Category of Hybrids and Hybrid Maps

3.3.1 Definition Let (X_1, Ω_1, R_1) and (X_2, Ω_2, R_2) be hybrids. A map $f : X_1 \to X_2$ is said to be **hybrid** if f is:

(1) continuous with respect to the topological components of these hybrids;
(2) strongly isotone with respect to their order components.

3.3.2 Lemma *Let* $(X_i, \Omega_i, R_i) \in \mathsf{HYB}$ *and* $\mathbf{C}_i$ *be the operation of topological closure in* (X_i, Ω_i) $(i = 1, 2)$. *The following are equivalent:*

(1) *the map* $f : X_1 \to X_2$ *is hybrid;*
(2) *for each* $x \in X_1$ *and each subset* A *of* X_1, *the following commutation conditions hold:*

$$\text{(a)}\ \ f\mathbf{C}_1(A) = \mathbf{C}_2 f(A) \quad \textit{and} \quad \text{(b)}\ \ f R_1(x) = R_2 f(x).$$

Proof Since each continuous map from a compact topological space to a Hausdorff space is closed, the continuity of the map f is equivalent to (2a), and by Proposition 1.4.12, f is strongly isotone if and only if (2b) holds. □

Let HYB (resp. HYB^+) denote the category of hybrids (resp. strict hybrids) and hybrid maps (see 3.1.9). Also recall that ST is the category of Stone spaces and continuous maps.

Our next task is to establish the duality of the category HYB of hybrids and the category CA of closure algebras, and that of the category HYB^+ of strict hybrids and the category HA of Heyting algebras.

3.3.3 Definition of Functors

(1a) To every closure algebra $(B, \mathbf{C})$ with the skeleton H we assign $(B, \mathbf{C})_* = (X, \Omega, R)$, where X is the set of all ultrafilters of the Boolean algebra B. The topology Ω is generated by the basis $\mathbb{F} = \{\varphi(a) : a \in B\}$ $(\varphi(a) = \{x \in X : a \in x\})$. The binary relation R is given by $xRy \Leftrightarrow a \in y$ implies $\mathbf{C}a \in x$ for each $a \in B$.
(1b) To every morphism $h : B_1 \to B_2$ of the category CA we assign the map $h_* = f : X_2 \to X_1$ given by $f(x) = h^{-1}(x)$, where x is an ultrafilter of B_2, $X_1 = (B_1)_*$, and $X_2 = (B_2)_*$.
(2a) To every hybrid $(X, \Omega, R) \in \mathsf{HYB}$ we assign the algebra $(B(X), \cup, \cap, -, \varnothing, X, R^{-1}) = (X, \Omega, R)^*$ of clopen sets, where $B(X)$ is the field of all clopen subsets of the Stone space X.
(2b) To every hybrid map $f : X_1 \to X_2$ $((X_i, \Omega_i, R_i) \in \mathsf{HYB}, i = 1, 2)$ we assign the map $f^* = h : B(X_2) \to B(X_1)$ given by $h(A) = f^{-1}(A)$ for each $A \in B(X_2)$.
(3) For every closure algebra $(B, \mathbf{C}) \in \mathsf{CA}$, let $\varphi_B = \varphi$ be the map $B \to B(X)$, where $X = B_*$, given by $\varphi(a) = \{x \in X : a \in X\}$ for $a \in B$.
(4) For every hybrid $(X, \Omega, R) \in \mathsf{HYB}$, let $\psi_X = \psi$ be the map $X \to B(X)_*$ given by $\psi(x) = \{A \in B(X) : x \in A\}$ for $x \in X$.

3.3.4 Duality Theorem *The correspondences* $(B, \mathbf{C}) \to (B, \mathbf{C})_*$, $h \to h_*$ *and* $(X, \Omega, R) \to (X, \Omega, R)^*$, $f \to f^*$ *are contravariant functors establishing the duality (= coequivalence) of the categories* CA *and* HYB.

The proof of the theorem is based on the following lemmas. Let $(B, \mathbf{C}) \in \mathsf{CA}$, H be its skeleton, and $(B, \mathbf{C})_* = (X, \Omega, R)$.

3.3.5 Lemma *The following are equivalent for $x, y \in X$:*

(1) *$a \in y$ implies $\mathbf{C}a \in x$ for each $a \in B$;*
(2) *$\mathbf{I}a \in x$ implies $a \in y$ for each $a \in B$;*
(3) *$x \cap H \subseteq y \cap H$.*

Proof (1 ⇔ 2) Condition (1) is equivalent to $(\forall a \in B)\,(\mathbf{C}a \notin x \Rightarrow a \notin y)$. Since x and y are ultrafilters, this condition is equivalent to $(\forall a \in B)\,(-\mathbf{C}a \in x \Rightarrow -a \in y)$. Replacing a by $-a$ and using $\mathbf{I}a = -\mathbf{C}{-a}$, we obtain the equivalent condition $(\forall a \in B)\,(\mathbf{I}a \in x \Rightarrow a \in y)$, which is condition (2).

(2 ⇒ 3) Let $a \in x \cap H$. Since $\mathbf{I}a = a$, by (2), $a \in y \cap H$.

(3 ⇒ 2) Let $x \cap H \subseteq y \cap H$ and $\mathbf{I}a \in x$. Then $\mathbf{I}a \in x \cap H$, so $\mathbf{I}a \in y \cap H$. Since $\mathbf{I}a \leq a$, we obtain $a \in y$. □

The following is an immediate consequence of Lemma 3.3.5.

3.3.6 Lemma *The relation R is reflexive and transitive, i.e., (X, R) is a quasi-ordered set.*

3.3.7 Lemma

(1) *The family $\mathbb{F} = \{\varphi(a) : a \in B\}$ is a basis of the space (X, Ω) consisting of clopen sets;*
(2) *the space (X, Ω) is compact, Hausdorff, and zero-dimensional;*
(3) *the map φ is a bijective isomorphism of the closure algebra $(B, \mathbf{C})$ onto the algebra of clopen sets $(B(X), R^{-1})$ of the hybrid (X, Ω, R).*

Proof (1) We have $\varphi(1) = X$ and $\varphi(a) \cap \varphi(b) = \varphi(a \wedge b)$ for $a, b \in B$ since $\varphi(a \wedge b) = \{x : a \wedge b \in x\} = \{x \in X : a \in x\} \cap \{x \in X : b \in x\}$, so $\mathbb{F}$ is a basis. It follows from the definition of the topology on X that every set $\varphi(a)$, where $a \in B$, is open. Moreover, $\varphi(-a) = X - \varphi(a)$ since $\varphi(-a) = \{x \in X : -a \in x\}$ $= \{x \in X : a \notin x\} = X - \varphi(a)$ because elements of X are ultrafilters. Thus, each $\varphi(a)$ is clopen.

(2) That X is zero-dimensional follows from (1). To see that X is Hausdorff, if $x, y \in X$ and $x \neq y$, then $x - y \neq \varnothing$, i.e., there exists $a \in x - y$. By (1), $\varphi(a)$ is clopen, $x \in \varphi(a)$, and $y \notin \varphi(a)$. To see that X is compact, let $K \subseteq B$, and suppose that $\bigcap\{\varphi(a) : a \in K\} = \varnothing$ but each finite subfamily has a nonempty intersection. Then $b_1, \ldots, b_k \in K$ implies $b_1 \wedge \cdots \wedge b_k \neq 0$, so the filter generated by K is proper. Hence, there exists an ultrafilter $x \in X$ such that $K \subseteq x$. Therefore, $x \in \bigcap\{\varphi(a) : a \in K\}$, a contradiction.

(3) It follows from the proof of (1) that $\varphi : B \to B(X)$ is a Boolean homomorphism, and reasoning as in the proof of (2) yields that φ is injective. We show that φ is surjective. If $A \in \mathbb{F}(X)$, then by (1), $A = \bigcup\{\varphi(a) : a \in K\}$ for some $K \subseteq B$. Since A is clopen and hence compact, there exists a finite subset $K_0 \subseteq K$ such that $A = \bigcup\{\varphi(a) : a \in K_0\}$. Therefore, $A = \varphi(b)$ for some $b \in B$. It remains to show that $\varphi(\mathbf{C}a) = R^{-1}(\varphi(a))$ for each $a \in B$. Recall that $a \in x$ is equivalent to $x \in \varphi(a)$. The equality $\varphi(\mathbf{C}a) = R^{-1}(\varphi(a))$ can be written as

$(\forall x \in X)(x \in \varphi(\mathbf{C}a) \Leftrightarrow x \in R^{-1}(\varphi(a)))$, which by Lemma 3.3.5 is equivalent to: for each $x \in X$, we have $\mathbf{C}a \in x \Leftrightarrow (\exists y \in X)\ (a \in y\ \&\ xRy)$. Let $a \in y$ and xRy, i.e., $(\forall b \in B)\ (b \in y \Rightarrow \mathbf{C}b \in x)$. Therefore, $\mathbf{C}a \in x$. Now let $\mathbf{C}a \in x$. We show that there is an ultrafilter $y \in X$ such that $a \in y$ and $(\forall b)\ (\mathbf{I}b \in x \Rightarrow b \in y)$. Consider the set $K_a = \{a\} \cup \{b : \mathbf{I}b \in x\}$. We claim that the filter generated by K_a is proper. Suppose $a \wedge b_1 \wedge \cdots \wedge b_k = 0$ where $\mathbf{I}b_i \in x$ for each b_i. Then $b_1 \wedge \cdots \wedge b_k \leq -a$ and hence $\mathbf{I}(b_1 \wedge \cdots \wedge b_k) \leq \mathbf{I}{-a}$. Therefore, $\mathbf{I}b_1 \wedge \cdots \wedge \mathbf{I}b_k \leq -\mathbf{C}a$. Since $\mathbf{I}b_1, \ldots, \mathbf{I}b_k \in x$ and x is a filter, it follows that $-\mathbf{C}a \in x$, which contradicts $\mathbf{C}a \in x$. Now extend K_a to an ultrafilter, say y. Clearly $a \in y$ and $K_a \subseteq y$. Let $\mathbf{I}b \in x$. Then $b \in K_a$, i.e., $b \in y$. The lemma is proved. □

3.3.8 Lemma *If A is a clopen subset of X, then $R^{-1}A$ is clopen.*

Proof Let A be clopen in X. Then, by Lemma 3.3.7(3), $A = \varphi(a)$ for some $a \in B$ and $R^{-1}(\varphi a) = \varphi(\mathbf{C}a)$. Thus, $R^{-1}(\varphi a)$ is clopen. □

As an immediate consequence of Lemma 3.3.7, we obtain the following.

3.3.9 Lemma *The algebra $(B(X), \cup, \cap, -, \varnothing, X, R^{-1})$, where $B(X)$ is the field of all clopen subsets of X, is a closure algebra isomorphic to $(B, \vee, \wedge, -, 0, 1, \mathbf{C})$.*

In what follows we call the algebra $(B(X), \cup, \cap, -, \varnothing, X, R^{-1})$ the **algebra of clopen sets** of (X, Ω, R).

3.3.10 Lemma *For each $x \in X$, the set $R(x)$ is closed.*

Proof Let $y \in X$ belong to the closure $\mathbf{C}R(x)$ of $R(x)$ in X. Then for each clopen set A, from $y \in A$ it follows that $R(x) \cap A \neq \varnothing$. We show that $y \in R(x)$. Suppose $y \notin R(x)$. By Lemmas 3.3.5 and 3.3.7, $y \notin R(x)$ is equivalent to the existence of a clopen set A such that $y \in A$ and $x \notin R^{-1}(A)$. Since $y \in R^{-1}(A)$ by the reflexivity of R, and $x \notin R^{-1}(A)$, we have $R^{-1}(A) \cap R(x) = \varnothing$. Indeed, if $z \in R^{-1}(A) \cap R(x)$, then $z \in R^{-1}(A)$, and as xRz, we have $x \in R^{-1}(A)$ by the transitivity of R, a contradiction. Since A is clopen, by Lemma 3.3.8, so is $R^{-1}(A)$. Therefore, there is a clopen set $A_0 = R^{-1}(A)$ such that $y \in A_0$ and $R(x) \cap A_0 = \varnothing$, i.e., $y \notin \mathbf{C}A_0$, a contradiction. Thus, $R(x)$ is closed for each $x \in X$. □

Lemmas 3.3.9 and 3.3.10 together yield the following.

3.3.11 Lemma *(X, Ω, R) is a hybrid, and the closure algebra $(B, \vee, \wedge, -, 0, 1, \mathbf{C})$ is isomorphic to the algebra of clopen sets $(B(X), \cup, \cap, -, \varnothing, X, R^{-1})$ of the hybrid (X, Ω, R).*

3.3.12 Lemma *Let (X, Ω, R) be an arbitrary hybrid and $\mathbb{F}$ a downward directed (with respect to inclusion) family of nonempty closed subsets of (X, Ω). Then*

$$R^{-1}\Big(\bigcap\{F : F \in \mathbb{F}\}\Big) = \bigcap\{R^{-1}F : F \in \mathbb{F}\}.^{11}$$

[11] *Editorial note*: This result is now known as **Esakia's lemma**; see, e.g., p. 350 of A. Chagrov and M. Zakharyaschev, *Modal Logic* (Clarendon Press), 1997. Note that the lemma does not require that R be reflexive or transitive.

Proof The equality in the lemma is equivalent to the following: for each $x \in X$,

$$x \in R^{-1}\Big(\bigcap\{F : F \in \mathbb{F}\}\Big) \Leftrightarrow (\forall F \in \mathbb{F})\ (x \in R^{-1}F)$$

or, in other words,

$$R(x) \cap \bigcap\{F : F \in \mathbb{F}\} \neq \varnothing \Leftrightarrow (\forall F \in \mathbb{F})\ (R(x) \cap F \neq \varnothing).$$

Obviously if $R(x) \cap \bigcap\{F : F \in \mathbb{F}\} \neq \varnothing$, then $R(x) \cap F \neq \varnothing$. Conversely, suppose $R(x) \cap F \neq \varnothing$ for each $F \in \mathbb{F}$. Since F is closed and $R(x)$ is closed for each $x \in X$ (Theorem 3.1.2(IIa)), the set $R(x) \cap F$ is closed. As $\mathbb{F}$ is downward directed, so is the family $\{R(x) \cap F : F \in \mathbb{F}\}$. Compactness of the space (X, Ω) implies that $\bigcap\{R(x) \cap F : F \in \mathbb{F}\}$ is nonempty, hence $R(x) \cap \bigcap\{F : F \in \mathbb{F}\} \neq \varnothing$. □

3.3.13 Lemma *Let* $(B_i, \mathbf{C}_i) \in \mathsf{CA}$, $h : B_1 \to B_2$ *be a morphism in the category* CA, (X_i, Ω_i, R_i) *be the hybrid corresponding to* $(B_i, \mathbf{C}_i)$ $(i = 1, 2)$, *and* $f = h_* = h^{-1}$. *Then:*

(1) f *maps the set* X_2 *to* X_1*;*
(2) f *is a continuous map from* (X_2, Ω_2) *to* (X_1, Ω_1)*;*
(3) f *is injective (resp. surjective) if and only if* h *is surjective (resp. injective);*
(4) f *is a strongly isotone map from* (X_2, R_2) *to* (X_1, R_1).

Proof (1) Let $x \in X_2$. We will show that $f(x) \in X_1$, i.e., that $f(x)$ is an ultrafilter of B_1. First let $a, b \in f(x) = h^{-1}(x)$, i.e., $ha, hb \in x$. Then $ha \wedge hb \in x$. But $ha \wedge hb = h(a \wedge b)$, i.e., $a \wedge b \in h^{-1}(x) = f(x)$. Next let $a \in f(x)$ and $a \leq b \in B_2$. Then $ha \in x$ and $ha \leq hb$. Therefore, $hb \in x$, i.e., $b \in h^{-1}(x) = f(x)$. Lastly let $a \vee b \in f(x)$, i.e., $h(a \vee b) \in x$. Then $ha \vee hb \in x$, and since x is an ultrafilter and hence a prime filter, $ha \in x$ or $hb \in x$, i.e., $a \in f(x)$ or $b \in f(x)$.

(2) We show that $f^{-1}(A)$ is open in (X_2, Ω_2) for each clopen subset A of X_1. Since $A = \varphi(a)$ for some $a \in B_1$,

$$\begin{aligned} f^{-1}(A) = f^{-1}(\varphi a) &= \{x \in X_2 : f(x) \in \varphi(a)\} = \{x \in X_2 : a \in f(x)\} \\ &= \{x \in X_2 : ha \in x\} = \varphi(ha) \in B(X_2). \end{aligned}$$

(3) Let f be injective. If h is not surjective, then there exists $a \in B_2 - h(B_1)$. As $h(B_1)$ is a subalgebra of B_2, by Makinson's lemma [11], there exist $x, y \in X_2$ such that $a \in x$, $a \notin y$, and $x \cap h(B_1) = y \cap h(B_1)$. Therefore, $x \neq y$ and $f(x) = f(y)$. Thus, f is not injective, a contradiction. Now let h be surjective. We show that $f(x) = f(y)$ implies $x = y$. Suppose $x \neq y$. Say $a \in x$ and $a \notin y$. Since h is surjective, $h^{-1}(a) \neq \varnothing$. Let $b \in h^{-1}(a)$. Then $hb = a$ and so $b \in f(x)$. But $hb = a \notin y$. Therefore, $b \notin f(y)$.

Next, let f be surjective. To show that h is injective, let $a, b \in B_1$ with $a \neq b$. Without loss of generality we may assume that $a \nleq b$. Therefore, there exists $x \in X_1$ such that $a \in x$ and $b \notin x$. Since f is surjective, there exists $y \in X_2$ such that $x = f(y)$. Thus, $ha \in y$ and $hb \notin y$, so $ha \neq hb$, and hence h is injective. Now let h be injective. We show that f is surjective, i.e., that for each $x \in X_1$ there exists $y \in X_2$ such that $f(y) = x$, i.e., $x = \{a \in B_1 : ha \in y\}$. Let F be the filter of B_2 generated by $h(x)$. Since h is injective, F is proper, so there is a $y \in X_2$ extending F. It is easy to check that $\{a \in B_1 : ha \in y\} = x$.

(4) We show that f is strongly isotone, i.e., that $f^{-1}R_1^{-1}(x) = R_2^{-1}f^{-1}(x)$ for $x \in X_1$ (see Proposition 1.4.12). Since X_1 is a Stone space and $\{\varphi(a) : a \in B_1\}$ is a basis of clopen sets, $\{x\} = \bigcap\{\varphi(a) : a \in x\}$, and $\{\varphi(a) : a \in x\}$ is a downward directed family of nonempty clopen sets in X_1. Therefore, the left-hand side of the equality can be rewritten as follows:

$$\begin{aligned} f^{-1}R_1^{-1}(x) &= f^{-1}R_1^{-1}\bigcap\{\varphi(a) : a \in x\} = f^{-1}\bigcap\{R_1^{-1}\varphi(a) : a \in x\} \\ &= f^{-1}\bigcap\{\varphi(\mathbf{C}_1 a) : a \in x\} = \bigcap\{f^{-1}\varphi(\mathbf{C}_1 a) : a \in x\} \\ &= \bigcap\{\varphi(h\mathbf{C}_1 a) : a \in x\}, \end{aligned}$$

where we have used Lemmas 3.3.12 and 3.3.7(3) and the proof of (2). The right-hand side can be rewritten as follows:

$$\begin{aligned} R_2^{-1}f^{-1}(x) &= R_2^{-1}f^{-1}\bigcap\{\varphi(a) : a \in x\} = R_2^{-1}\bigcap\{f^{-1}\varphi(a) : a \in x\} \\ &= R_2^{-1}\bigcap\{\varphi(ha) : a \in x\} = \bigcap\{R_2^{-1}\varphi(ha) : a \in x\} \\ &= \bigcap\{\varphi(\mathbf{C}_2 ha) : a \in x\}, \end{aligned}$$

where again we have used Lemmas 3.3.12 and 3.3.7(3) and the proof of (2). Since h is a closure algebra homomorphism, $h\mathbf{C}_1 a = \mathbf{C}_2 ha$. Thus, $f^{-1}R_1^{-1}(x) = R_2^{-1}f^{-1}(x)$. □

3.3.14 Lemma *Let $(X_i, \Omega_i, R_i) \in \mathsf{HYB}$, $(B_i, \mathbf{C}_i) = (X_i, \Omega_i, R_i)^*$ be the corresponding closure algebra $(i = 1, 2)$,*[12] *$f : X_1 \to X_2$ a continuous strongly isotone map, and $h = f^* = f^{-1}$. Then:*

(1) *h maps B_2 to B_1;*
(2) *h is a morphism of the category* CA*;*
(3) *h is injective (resp. surjective) if and only if f is surjective (resp. injective).*

Proof (1) Since the f-preimage of a clopen set is clopen, $h = f^{-1}$ maps B_2 to B_1.

(2) Since $f^{-1}(A \cup B) = f^{-1}(A) \cup f^{-1}(B)$ and $f^{-1}(-A) = -f^{-1}(A)$, we have that h is a Boolean homomorphism. We prove that h preserves the closure operator, i.e., that $f^{-1}(R_2^{-1}A) = R_1^{-1}f^{-1}(A)$ for each clopen set A of (X_2, Ω_2). This equation is equivalent to $\bigcup\{f^{-1}R_2^{-1}(x) : x \in A\} = \bigcup\{R_1^{-1}f^{-1}(x) : x \in A\}$. The latter equation follows from the fact that f is strongly isotone (see Proposition 1.4.12).

(3) Easy to verify.[13] □

---

[12] *Editorial note*: It is straightforward to see that $(X_i, \Omega_i, R_i)^* = (B(X_i), R_i^{-1})$ is indeed a closure algebra.

[13] *Editorial note*: Indeed, it is easy to see that f being surjective implies that f^{-1} is injective. For the converse, given $y \in Y$, we have that $\{y\}$ is the intersection of the downward directed family of

3.3.15 Lemma *The map* $\psi : X \to B(X)_*$ *is:*

(1) *a homeomorphism of the spaces* (X, Ω) *and* $(X_0, \Omega_0) = (X, \Omega)^*{}_*$;
(2) *an isomorphism of the quasi-ordered sets* (X, R) *and* $(X_0, R_0) = (X, R)^*{}_*$.

Proof (1) First we show that ψ is surjective, i.e., that $x_0 \in X_0 = X^*{}_*$ implies $\psi^{-1}(x_0) \neq \varnothing$. Since x_0 is an ultrafilter of $B(X) = X^*$, the family $\{A \in B(X) : A \in x_0\}$ of clopen sets has the finite intersection property. Then since X is compact by Lemma 3.3.7(2), there exists x such that $x \in \bigcap\{A \in B(X) : A \in x_0\}$. But then $x_0 = \psi(x)$. Next we show that ψ is injective. Indeed, if $x \neq y$, then since X is Hausdorff and zero-dimensional by Lemma 3.3.7(2), we have $\psi(x) \neq \psi(y)$. Finally, ψ is continuous, i.e., the ψ-preimage of each element of a basis of $B(X)_*$ is open, i.e., $\psi^{-1}(\varphi A)$ is open for $A \in B(X)$ where $\varphi(A) = \varphi_{B(X)}(A)$. Indeed, $y \in \psi^{-1}(\varphi A) \Leftrightarrow \psi(y) \in \varphi(A) \Leftrightarrow A \in \psi(y) \Leftrightarrow y \in A$. Therefore, $\psi^{-1}(\varphi A) = A$, and the continuity of ψ has been proved. Since X and $X^*{}_*$ are compact Hausdorff spaces, ψ is a homeomorphism.

(2) We show that ψ is an isomorphism of the quasi-ordered sets (X, R) and $(X_0, R_0) = (X, R)^*{}_*$. Note first that if xRy, then for each clopen subset A of (X, Ω), from $y \in A$ it follows that $x \in R^{-1}(A)$. The converse is also true. Indeed, if $y \notin R(x)$, then $y \in -R(x)$. Since $R(x)$ is closed, $-R(x)$ is open. As (X, Ω) is zero-dimensional, there exists a clopen set A such that $y \in A \subseteq -R(x)$, so $x \notin R^{-1}A$. Therefore, xRy is equivalent to the following: $y \in A$ implies $x \in R^{-1}A$ for each clopen A. (From this, in particular, it follows that if (X, Ω, R_1) and (X, Ω, R_2) are hybrids with the same topological component (X, Ω), and $R_1^{-1}(A) = R_2^{-1}(A)$ for each clopen A, then $R_1 = R_2$.)

We are ready to show that $xRy \Leftrightarrow \psi(x)R_0\psi(y)$. By the definition of the relation R_0 on X_0, we have: $\psi(x)R_0\psi(y) \Leftrightarrow (A \in \psi(y) \Rightarrow R^{-1}A \in \psi(x))$ for each clopen $A \in B(X)$. By the definition of ψ, the right-hand side is equivalent to the implication $y \in A \Rightarrow x \in R^{-1}(A)$ for each $A \in B(X)$. This, in turn, is equivalent to xRy by the previous paragraph. Therefore, $xRy \Leftrightarrow \psi(x)R_0\psi(y)$. Since ψ is a bijection, ψ is an isomorphism by Proposition 1.4.15. The lemma is proved. □

The Duality Theorem is now proved.[14]

Bibliographic notes. The Duality Theorem was first published in [4]. See also [5].

---

clopens A_i containing y. Since each A_i is nonempty, and f^{-1} is injective, $f^{-1}(A_i)$ is nonempty. Thus, by compactness, the intersection of the $f^{-1}(A_i)$ is nonempty, so $f^{-1}(\{y\})$ is nonempty. This shows that f is surjective.

To see that f^{-1} being surjective implies that f is injective, use the fact that distinct points are distinguished by clopens. For the converse, suppose A is a clopen subset of X. Then $f(A)$ is closed in Y, so $f(A)$ is the intersection of clopens B_i such that $f(A) \subseteq B_i$. Thus, $f^{-1}(f(A)) = \bigcap\{f^{-1}(B_i) \mid i \in I\}$, and since f is injective, $A = f^{-1}(f(A))$, so $A = \bigcap\{f^{-1}(B_i) \mid i \in I\}$. Using compactness, it follows that there is one clopen B such that $A = f^{-1}(B)$.

Note that Lemmas 3.3.13(3) and 3.3.14(3) are inter-derivable using the Duality Theorem (the proof of which does not require these lemmas).

[14] *Editorial note*: That $(\cdot)_*$ is a well-defined contravariant functor follows from Lemmas 3.3.11 and 3.3.13 (and the easily verifiable facts that it preserves identity and composition), and that $(\cdot)^*$ is a well-defined contravariant functor follows from Lemma 3.3.14 (and the same easily verifiable facts). That $(\cdot)_*$ satisfies part (a) of the definition of coequivalence (Definition 1.2.2(2)) follows from Lemma 3.3.15. That $(\cdot)_*$ satisfies part (b) is a consequence of Lemmas 3.3.13 and 3.3.14

3.4 The Category of Heyting Algebras and the Category of Strict Hybrids

Recall that a ring of cones over a quasi-ordered set is, by definition, a ring of sets $\mathbb{K}$ such that $A_1, A_2 \in \mathbb{K}$ imply $\{x : R(x) \cap (A_1 - A_2) = \varnothing\} \in \mathbb{K}$ (Definition 2.2.9).

Let (X, Ω, R) be a hybrid and $(B(X), \cup, \cap, -, \varnothing, X, R^{-1})$ the algebra of clopen sets of the hybrid (X, Ω, R) (Lemma 3.3.9). Let $\mathbb{H}(X)$ denote the set of all clopen cones of (X, Ω, R).

3.4.1 Proposition

(1) *The set* $\mathbb{H}(X)$ *is a ring of cones over the quasi-ordered set* (X, R)*;*
(2) $A \in \mathbb{H}(X)$ *if and only if* $A = -R^{-1}(A_0)$ *for some clopen set* $A_0 \in B(X)$.

Proof (1) Obviously $\varnothing, X \in \mathbb{H}(X)$, and if $A_1, A_2 \in \mathbb{H}(X)$, then $A_1 \cup A_2$, $A_1 \cap A_2 \in \mathbb{H}(X)$. Let $A_1, A_2 \in \mathbb{H}(X)$. We will show that the set $\{x \in X : R(x) \cap (A_1 - A_2) = \varnothing\}$ belongs to $\mathbb{H}(X)$. Since A_1 and A_2 are clopen sets of (X, Ω), $A_1 - A_2$ is clopen, and hence $R^{-1}(A_1 - A_2)$ is clopen by Theorem 3.1.2(IIIb). Since $R^{-1}(A_1 - A_2)$ is a lower cone, $-R^{-1}(A_1 - A_2)$ is a clopen upper cone, and so $-R^{-1}(A_1 - A_2) \in \mathbb{H}(X)$. As $-R^{-1}(A_1 - A_2) = \{x \in X : R(x) \cap (A_1 - A_2) = \varnothing\}$, the set $\{x : R(x) \cap (A_1 - A_2) = \varnothing\}$ belongs to $\mathbb{H}(X)$.

(2) follows from the proof of (1). □

3.4.2 Corollary *Let* (X, Ω, R) *be a hybrid. Then the set* $\mathbb{H}(X)$ *of all clopen cones of* (X, Ω, R) *forms a Heyting algebra which is the skeleton of the algebra of clopen sets* $(B(X), R^{-1})$ *of the hybrid.*

Proof By Proposition 3.4.1(1), $(\mathbb{H}(X), \cup, \cap, \rightarrow, \varnothing, X)$ is a Heyting algebra, where $A_1 \rightarrow A_2 = \{x : R(x) \cap (A_1 - A_2) = \varnothing\}$ for $A_1, A_2 \in \mathbb{H}(X)$. By Proposition 3.4.1 (2), $\mathbb{H}(X)$ is the skeleton of the algebra. □

We call the Heyting algebra $\mathbb{H}(X)$ **the ring of clopen cones** of the hybrid (X, Ω, R).

3.4.3 Convention Based on the Duality Theorem (Theorem 3.3.4), we will often conveniently identify the closure algebra $(B, \mathbf{C})$ with the algebra of clopen sets $(B(X), R^{-1})$ of the corresponding hybrid (X, Ω, R) and its skeleton H with the ring of clopen cones $\mathbb{H}(X)$ of the hybrid.

3.4.4 Theorem *A closure algebra* $(B, \mathbf{C})$ *is a skeletal closure algebra if and only if the corresponding hybrid* (X, Ω, R) *is strict (i.e.,* (X, R) *is a partially ordered set).*

Proof Use Convention 3.4.3 and identify $(B, \mathbf{C})$ with $(B(X), R^{-1})$ and its skeleton H with $\mathbb{H}(X)$. Let $(B(X), R^{-1})$ be a skeletal closure algebra. Then the field of clopen sets $B(X)$ is the smallest field containing the ring of clopen cones $\mathbb{H}(X)$. In other words, each clopen $A \in B(X)$ can be obtained from finitely many clopen cones from $\mathbb{H}(X)$ by applying the set-theoretic operations of union $\cup$, intersection $\cap$, and complement $-$. To show that R is anti-symmetric, i.e., that there are no nontrivial clusters in (X, R), suppose there are $x, y \in X$ such that $x \neq y$, xRy, and yRx. Since $x \neq y$, there exists a clopen set A such that $x \in A$ and $y \notin A$. Clearly A cannot be either an upper or a lower cone since xRy and yRx. The clopen A cannot be obtained by applying $\cup$, $\cap$, and $-$ to a finite number of clopen cones because every clopen cone A_0 has the property that either $x, y \in A_0$ or $x, y \notin A_0$, and the operations $\cup$, $\cap$, $-$ preserve this property. So $A \notin B(X)$, contradicting the fact that $(B, \mathbf{C})$ is a skeletal algebra.

Now let (X, R) be a partially ordered set. Then for each $x, y \in X$, from $x \neq y$ it follows that either $x \notin R(y)$ or $y \notin R(x)$. Without loss of generality we may assume that $x \notin R(y)$. Then $x \in -R(y)$. Since $R(y)$ is closed in the space (X, Ω), the set $-R(y)$ is open. As (X, Ω) is zero-dimensional, there exists a clopen set A such that $x \in A \subseteq -R(y)$. Since $-R(y)$ is a lower cone, $R^{-1}(A) \subseteq -R(y)$. Moreover, $x \in R^{-1}(A)$, $y \notin R^{-1}(A)$, and as A is clopen, by Theorem 3.1.2(IIIb), $R^{-1}(A)$ is clopen. Therefore, $x \notin -R^{-1}(A)$ and $y \in -R^{-1}(A)$. Thus, the ring of clopen cones separates any two distinct points of the space (X, Ω), i.e., it is reduced. Hence, the field $\mathbb{F}$ generated by it is reduced. Since (X, Ω) is a compact Hausdorff space and $\mathbb{F}$ is a reduced field of clopen sets, by Proposition 1.3.2, the field coincides with the field $B(X)$ of all clopen sets of (X, Ω, R). □

3.4.5 Construction Let $(B, \mathbf{C})$ be a skeletal closure algebra, (X, Ω, R) the corresponding strict hybrid, and H the skeleton of the algebra. Let X_0 denote the set of all prime filters of the Heyting algebra H. Define a relation R_0 on X_0 by $x R_0 y \Leftrightarrow x \subseteq y$ for each $x, y \in X_0$. Introduce a topology Ω_0 on the set by declaring the family $\mathbb{F}_0 = \{\varphi(a) : a \in H\} \cup \{-\varphi(a) : a \in H\}$, where $\varphi(a) = \{x \in X_0 : a \in x\}$, as a subbasis of open sets. [15]

That the two constructions coincide is stated in the following.

3.4.6 Theorem

(1) *The space* (X, Ω) *is homeomorphic to the space* (X_0, Ω_0).
(2) *The partially ordered set* (X, R) *is isomorphic to the partially ordered set* (X_0, R_0).

Proof Define a map $f : X \to X_0$ by setting $f(x) = x \cap H$ for each $x \in X$.

(1) To show that f is injective, suppose $f(x) = f(y)$, i.e., $x \cap H = y \cap H$, and $x \neq y$. From $x \neq y$ it follows that there is $a \in B$ such that $a \in x$ and $a \notin y$. Since x, y are ultrafilters, $-a \notin x$ and $-a \in y$. As $(B, \mathbf{C})$ is a skeletal algebra, $a = \bigvee_{k<m} (a_k \wedge -b_k)$ for some $m \in \omega$, $a_k, b_k \in H$. Since x is an ultrafilter and $a \in x$, $a_{k_0} - b_{k_0} \in x$ for some $k_0 < m$, i.e. $a_{k_0} \in x$ and $b_{k_0} \notin x$. On the other hand, $-a = \bigwedge_{k<m} -(a_k \wedge -b_k) \in y$, and as y is an ultrafilter, for all $k < m$, we have $-(a_k \wedge -b_k) \in y$, i.e., $a_k \notin y$ or $b_k \in y$. So $a_{k_0} \in x$, $b_{k_0} \notin x$, and $(\forall k < m)$ $(a_k \in y \Rightarrow b_k \in y)$. Since $a_k, b_k \in H$ for all $k < m$, $x \cap H = y \cap H$, and $a_{k_0} \in x$,

[15] *Editorial note*: It is easy to see that (X_0, Ω_0) is a zero-dimensional Hausdorff space.

we have $a_{k_0} \in y$. Therefore, $b_{k_0} \in y$, and as $x \cap H = y \cap H$, we arrive at a contradiction: $b_{k_0} \in x$ and $b_{k_0} \notin x$. Thus, f is injective.

Next we show that f is surjective. Let p be a prime filter of the Heyting algebra H. To prove that there exists an ultrafilter x of B such that $f(x) = p$, consider the family $\mathbb{F}_p$ of all proper filters F of the Boolean algebra B such that $F \cap H = p$. The family $\mathbb{F}_p$ is nonempty because the skeletal filter

$$F_p = \{a \in B : (\exists b \in p)\,(b \le a)\}$$

generated by the prime filter p (see Proposition 2.4.12) is proper, and $F_p \cap H = p$, so $F_p \in \mathbb{F}_p$. Let $\mathbb{K}$ be a nonempty chain in $\mathbb{F}_p$. Then $F_0 = \bigcup\{F : F \in \mathbb{K}\}$ is a proper filter of the Boolean algebra B, and $F_0 \cap H = p$ since $\bigcup\{F : F \in \mathbb{K}\} \cap H = \bigcup\{F \cap H : F \in \mathbb{K}\} = p$. Hence, by Zorn's lemma,[16] $\mathbb{F}_p$ has a maximal element, say x. To show that x is an ultrafilter of the Boolean algebra B, suppose $a \vee b \in x$ and $a, b \notin x$ for some $a, b \in B$. Since x is a maximal element of F_p, we have the following for the filter $F_a = [x, a]$ (resp. $F_b = [x, b]$) generated by the filter x and the element a (resp. b) : $(F_a \cap H) - p \neq \varnothing$(resp. $(F_b \cap H) - p \neq \varnothing$). Let $d_1 \in (F_a \cap H) - p$ and $d_2 \in (F_b \cap H) - p$. Then $a \wedge a_1 \le d_1$ and $b \wedge a_2 \le d_2$ for some $a_1, a_2 \in x$. Therefore, $(a \wedge a_1) \vee (b \wedge a_2) \le d_1 \vee d_2$. But

$$(a \wedge a_1) \vee (b \wedge a_2) = (a \vee b) \wedge (a \vee a_2) \wedge (a_1 \vee b) \wedge (a_1 \vee a_2).$$

We have $a \vee b \in x$; $a \vee a_2 \in x$ since $a_2 \in x$; $a_1 \vee b \in x$ since $a_1 \in x$; and $a_1 \vee a_2 \in x$. Thus, $d_1 \vee d_2 \in x$. But $d_1, d_2 \in H$, so $d_1 \vee d_2 \in p$, a contradiction. Consequently, x is an ultrafilter of B, and f is surjective.

Now we show that the bijection f is continuous. Let A be a subbasic clopen subset of X_0. Then either (I) $A = \{p \in X_0 : a \in p\}$, where $a \in H$, or (II) $A = \{p \in X_0 : a \notin p\}$.

Case I: We have $f^{-1}(A) = \{x \in X : f(x) \in A\} = \{x \in X : a \in f(x)\} = \{x \in X : a \in x \cap H\}$. But since $a \in H$, we have $\{x \in X : a \in x \cap H\} = \{x : a \in x\}$, and $\{x \in X : a \in x\}$ is clopen in X.

Case II: $f^{-1}(A) = \{x \in X : f(x) \in A\} = \{x : f(x) \in \{p \in X_0 : a \notin p\}\} = \{x \in X : a \notin f(x)\} = \{x \in X : a \notin x \cap H\}$. But as $a \in H$, we have $\{x \in X : a \notin x \cap H\} = \{x \in X : a \notin x\}$, and $\{x \in X : -a \in x\}$ is clopen in X. Since f is a continuous bijection of the compact space X onto the Hausdorff space X_0, f is a homeomorphism.

(2) We show that the bijection f is an isomorphism of (X, R) onto (X_0, R_0). By Proposition 1.4.15, it is sufficient to prove that $xRy \Leftrightarrow f(x)R_0f(y)$, i.e., that $xRy \Leftrightarrow x \cap H \subseteq y \cap H$. But the latter follows from Lemma 3.3.5. The lemma is proved. □

3.4.7 Remark Together the Duality Theorem (Theorem 3.3.4) and Theorem 3.4.6 establish Theorem 2.5.11, stating that the category HA of Heyting algebras and the category SA of skeletal closure algebras are equivalent.

---

16 *Editorial note*: Zorn's lemma states that a partially ordered set in which every chain has an upper bound has a maximal element (see, e.g., [11]).

We now have the following important corollary.

3.4.8 Corollary

(1) *The category* SA *of skeletal closure algebras is dual to the category* HYB^+ *of strict hybrids.*
(2) *The category* HA *of Heyting algebras is dual to the category* HYB^+ *of strict hybrids.*

Proof (1) Theorems 3.3.4 and 3.4.4.
(2) Theorems 3.3.4, 3.4.4, and 3.4.6. □

3.4.9 Corollary *Let H be a Heyting algebra and (X, Ω, R) the corresponding strict hybrid. Then H is a Boolean algebra if and only if R is the identity relation.*

Using Theorem 3.3.4, we prove the following useful lemma generalizing a fact known for Boolean algebras.

3.4.10 Lemma *Let H be a Heyting algebra, H_0 a subalgebra of H, and $a \in H$. The following are equivalent:*

(1) $a \notin H_0$;
(2) *there exist prime filters x, x' of H such that $a \in x$, $a \notin x'$, and $x \cap H_0 = x' \cap H_0$.*

Proof ($2 \Rightarrow 1$) is obvious. We prove ($1 \Rightarrow 2$). Let H_0 be a subalgebra of H, $h : H_0 \to H$ the inclusion map, and $f = h_* : X \to X_0$ the map of the corresponding hybrids (X, Ω, R) and (X_0, Ω_0, R_0) dual to h. Since h is injective, f is a surjective hybrid map.[17] Consider the set $f(\varphi(a)) = A_0 \subseteq X_0$, where $\varphi(a) = \{x \in X : a \in x\}$ is a clopen cone of (X, R). Since f is a hybrid map, f is continuous, and so f is a closed map preserving cones (Proposition 1.4.12). Hence, $f(\varphi(a)) = A_0$ is a closed cone. If $f^{-1}(A_0) = \varphi(a)$, then (as f is surjective) $f(X - \varphi(a)) = X_0 - A_0$. Since $X - \varphi(a)$ is clopen, $X_0 - A_0$ is closed in the space (X_0, Ω_0). Therefore, A_0 is a clopen cone, and so $A_0 = \varphi(a_0)$ for some $a_0 \in H_0$. Thus, $h(a_0) = a$, a contradiction. It follows that $\varphi(a) \subseteq f^{-1}(A_0)$ and $f^{-1}(A_0) - \varphi(a) \neq \varnothing$, i.e., there exists $x' \in f^{-1}(A_0)$ such that $x' \notin \varphi(a)$. But $f(x') \in A_0 = f(\varphi(a))$, so there exists $x \in \varphi(a)$ such that $f(x) = f(x')$; x and x' are the prime filters we sought. □

3.4.11 Lemma *Let (X, Ω, R) be a hybrid and X_0 a closed cone of the hybrid. Let Ω_0 be the subspace topology on X_0 and R_0 the restriction of R to X_0, i.e., $R_0(x) = R(x) \cap X_0$ for all $x \in X_0$. Then (X_0, Ω_0, R_0) is a hybrid.*

Proof Since X_0 is closed in (X, R), (X_0, Ω_0) is a Stone space. Clearly (X_0, R_0) is a quasi-ordered set. For each $x \in X$, $R_0(x) = R(x) \cap X_0$. Hence, $R_0(x)$ is closed in (X_0, Ω_0). So (IIa) of Theorem 3.1.2 holds. Let $A_0 = A \cap X_0$, where A is an arbitrary subset of X. We have

[17] *Editorial note*: The proof is similar to that of the analogous part of Lemma 3.3.13(3).

$$\begin{aligned}
R_0^{-1}(A_0) &= R_0^{-1}(A \cap X_0) = \bigcup\{R_0^{-1}(x) : x \in A \cap X_0\} \\
&= \bigcup\{R^{-1}(x) \cap X_0 : x \in A \cap X_0\} \\
&= X_0 \cap \bigcup\{R^{-1}(x) : x \in A \cap X_0\} = X_0 \cap R^{-1}(A \cap X_0) \\
&= X_0 \cap R^{-1}(A).
\end{aligned}$$

Notice that the fact that X_0 is a cone is needed only for the last equality. Indeed, we show that $X_0 \cap R^{-1}(A) \subseteq X_0 \cap R^{-1}(A \cap X_0)$. The reverse inclusion is obvious. Let $x \in X_0 \cap R^{-1}(A)$. Then $x \in X_0$ and $R(x) \cap A \neq \varnothing$. Let $y \in R(x) \cap A$. Then xRy. Since $x \in X_0$ and X_0 is a cone, $y \in X_0$. Therefore, $y \in R(x) \cap A \cap X_0$, so $R(x) \cap A \cap X_0 \neq \varnothing$. Thus, $x \in X_0 \cap R^{-1}(A \cap X_0)$. Now let A_0 be a closed subset of the space (X_0, Ω_0). Then since X_0 is closed in (X, Ω), $A_0 = A \cap X$ for some set A closed in (X, Ω). But $R_0^{-1}(A_0) = X_0 \cap R^{-1}(A)$, and since $R^{-1}(A)$ is closed in (X, Ω), the set $R_0^{-1}(A_0)$ is closed in (X_0, Ω_0). A similar argument shows that if U_0 is open in X_0, then $R_0^{-1}(U_0)$ is open in X_0. Consequently, condition (IIa) of Theorem 3.1.2 holds, and so (X_0, Ω_0, R_0) is a hybrid. □

3.4.12 Definition The hybrid (X_0, Ω_0, R_0), where X_0 is a closed cone of a hybrid (X, Ω, R), is called a **subhybrid** of (X, Ω, R).

Recall that if (X, Ω) is an arbitrary topological space, where Ω is the family of all its open sets, E is an equivalence relation on the set X, and $f : X \to X/E$ is the canonical map ($f(x) = E(x),\ x \in X$), then the set $X/_E$ equipped with the quotient topology $\Omega_E = \{A \in X/E : f^{-1}(A) \in \Omega\}$ is said to be a **quotient space** of (X, Ω).

3.4.13 Lemma *Let (X, Ω, R) be a hybrid, $(X_E, R/E)$ the skeleton of the quasi-ordered set (X, R), and $(X/E, \Omega_E)$ the quotient space. Then $(X/E, \Omega_E, R_E)$ is a strict hybrid, and the canonical map $f : X \to X/E$ is a hybrid map.*

Proof Obviously f is a continuous map from (X, Ω) onto $(X/E, \Omega_E)$ and, by Proposition 1.4.13, f is a strongly isotone map from (X, R) onto $(X/E, R_E)$. Note that since $(X/E, R_E)$ is the skeleton of (X, R), it follows that $(X/E, R_E)$ is a partially ordered set (see the discussion after Definition 1.4.7).

It remains to show that $(X/E, \Omega_E, R_E)$ is a hybrid. Clearly $(X/E, \Omega_E)$ is compact. We show that it is Hausdorff and zero-dimensional. Let $f(x)$ and $f(y)$ be two distinct points of X/E. Then $E(x) \neq E(y)$, i.e., $x \notin E(y)$. Thus, either $y \notin R(x)$ or $x \notin R(y)$. Without loss of generality we may assume that $y \notin R(x)$. By Theorem 3.1.2(IVa), there is a partition of X into two open sets A_1 and A_2 such that $x \in A_1$, $y \in A_2$, A_1 is an upper cone, and A_2 is a lower cone. Clearly $f(x) \in f(A_1)$ and $f(y) \in f(A_2)$, i.e., $E(x) \subseteq E(A_1)$ and $E(y) \subseteq E(A_2)$. Moreover, $E(y) \nsubseteq E(A_1)$ and $E(x) \nsubseteq E(A_2)$ since A_1 and A_2 are cones. For the same reason, $E(A_1) = A_1$ and $E(A_2) = A_2$. Thus, $f(x)$ and $f(y)$ are separated by disjoint clopen sets $f(A_1)$ and $f(A_2)$.[18] So $(X/E, \Omega_E)$ is a Stone space.

[18] *Editorial note*: Since $(X/E, \Omega_E)$ is compact and distinct points are separated by clopens, it follows that $(X/E, \Omega_E)$ is zero-dimensional. See, e.g., p. 69 of P. Johnstone, *Stone Spaces* (Cambridge University Press), 1982.

Now let $E(x) = f(x)$ be a point in X/E. We show that $R_E(E(x))$ is closed in $(X/E, \Omega_E)$. We have

$$R_E(E(x)) = \{E(y) : E(x)R_E E(y)\} = \{E(y) : y \in R(x)\}.$$

But $E(R(x)) = \bigcup\{E(y) : y \in R(x)\} = R(R(x) \cap R^{-1}(x))$. Since the latter set is closed in (X, Ω), the set $\{E(y) : y \in R(x)\} = R_E(E(x))$ is closed by the definition of the quotient topology. Finally, let $\mathbb{K} = \{E(x) : x \in A\}$ be a clopen set in $(X/E, \Omega_E)$. We show that $R_E^{-1}(\mathbb{K})$ is clopen. We have

$$\begin{aligned} R_E^{-1}(\mathbb{K}) &= \bigcup\{R_E^{-1}(E(x)) : x \in A\} \\ &= \bigcup\{\{E(y) : E(y) \in R_E^{-1}(E(x))\} : x \in A\} \\ &= \bigcup\{E(y) : y \in R^{-1}(A)\} = E(R^{-1}(A)) = R^{-1}(A). \end{aligned}$$

Since $\mathbb{K}$ is clopen in X/E, by the definition of the quotient topology we may assume that A is clopen in X. Then $R^{-1}(A)$ is clopen in X by Theorem 3.1.2(IIIb). Therefore, $R_E^{-1}(\mathbb{K})$ is clopen in X/E. Thus, $(X/E, \Omega_E, R_E)$ is a hybrid, and the canonical map $f : X \to X/E$ is a hybrid map. □

3.4.14 Definition If (X, Ω, R) is a hybrid and E is the equivalence relation on X defined by

$$xEy \Leftrightarrow xRy \ \& \ yRx,$$

then the strict hybrid $(X/E, \Omega_E, R_E)$ is called the **skeleton** of the hybrid (X, Ω, R).

3.4.15 Proposition *Let $(X/E, \Omega_E, R_E)$ be the skeleton of a hybrid (X, Ω, R), $(B(X), R^{-1})$ and $(B(X/E), R_E^{-1})$ the corresponding algebras of clopen sets, and $\mathbb{H}(X)$ and $\mathbb{H}(X/E)$ the corresponding algebras of clopen cones of these hybrids. Then:*

(1) *the closure algebra $(B(X/E), R_E^{-1})$ is embeddable in the closure algebra $(B(X), R^{-1})$;*
(2) *the algebra $(B(X/E), R_E^{-1})$ is a skeletal algebra;*
(3) *the Heyting algebras $\mathbb{H}(X)$ and $\mathbb{H}(X/E)$ are isomorphic.*

Proof (1) Since the canonical map $f : X \to X/E$ is surjective, the dual hybrid map $h = f^{-1} : B(X/E) \to B(X)$ is an embedding by Lemma 3.3.14.

(2) That $B(X/E)$ is a skeletal algebra follows from Theorem 3.4.4.

(3) For the map h from (1), we have that $A \in \mathbb{H}(X/E)$ implies $h(A) \in \mathbb{H}(X)$. Let $h' : \mathbb{H}(X/E) \to \mathbb{H}(X)$ be the restriction of h to $\mathbb{H}(X/E)$. Then h' is the desired isomorphism. □

Let $(B, \mathbf{C})$ be a closure algebra, H its skeleton, and $\Theta(H)$ (resp. $\Theta(B, \mathbf{C})$) the lattice of congruences of the Heyting algebra H (resp. of the closure algebra $(B, \mathbf{C})$). Let (X, Ω, R) be the hybrid corresponding to $(B, \mathbf{C})$. Let $\Xi(X, \Omega, R)$ (or simply ΞX) denote the family of all closed cones of (X, Ω, R) (recall Definition 3.4.12).

3.4.16 Theorem *The following pairs of lattices are dually isomorphic:*

(1) $\Theta(B, \mathbf{C})$ *and* $(\Xi X, \subseteq)$*;*
(2) $\Theta(H)$ *and* $(\Xi X, \subseteq)$*;*
(3) $\Theta(H)$ *and* $(\Xi(X/E), \subseteq)$*, where* $(X/E, \Omega_E, R_E)$ *is the skeleton of the hybrid* (X, Ω, R)*.*

Proof By Theorem 2.4.17, $\mathscr{F}(H) \cong \mathscr{F}^s(B, \mathbf{C})$ (see Definition 2.4.13). By Proposition 2.4.9(b), $\Theta(H) \cong \mathscr{F}(H)$. By Corollary 2.4.20, $\Theta(H) \cong \Theta(B, \mathbf{C})$. Note that $\Xi X \cong \Xi(X/E)$ since the existence of clusters (irrespective of their cardinality) does not affect the "stock" of cones. Indeed, if a cone A contains a point x, then $E(x) \subseteq A$. Assign to a skeletal filter F of $(B, \mathbf{C})$ the set A_F in X as follows: $A_F = \bigcap\{\varphi(\mathbf{I}a) : a \in F\}$, where φ is the isomorphism of B onto the algebra $B(X)$ of clopen sets of the hybrid X, i.e., $\varphi(a) = \{x \in X : a \in x\}$ for $a \in B$. We have $\varphi(\mathbf{I}a) = -R^{-1}-\varphi(a)$ (recall Lemma 3.3.7(3)). Since $\varphi(a)$ is clopen, $-R^{-1}-\varphi(a)$ is a clopen cone. Clearly A_F is a closed cone (being an intersection of clopen cones), that is, $A_F \in \Xi X$. Obviously $F \subseteq F'$ implies $A_{F'} \subseteq A_F$. It is easy to see that the indicated correspondence establishes the required dual isomorphism. □

Recall that a principal cone (i.e., a cone of the form $R(x)$, where $x \in X$) is closed. Note that the above correspondence establishes also a bijection between the set of prime skeletal filters of $(B, \mathbf{C})$ (as well as the set of prime filters of the Heyting algebra H) and the set of principal cones.

As byproducts of Theorem 3.3.4 (and Corollary 3.4.8), we obtain:

3.4.17 Theorem

(1) *The category* POF *of all finite partially ordered sets is dual to the category* FHA *of all finite Heyting algebras.*
(2) *The category* QOF *of all finite quasi-ordered sets is dual to the category* FCA *of all finite closure algebras.*

3.4.18 Representation Theorem *Each Heyting algebra is isomorphic to a ring of cones of a suitable partially ordered set.*

3.4.19 Corollary *Each finite Heyting algebra (and so each finite distributive lattice) is isomorphic to the ring of* ***all*** *cones of some partially ordered set.*

Bibliographic notes. Lemma 3.4.10 extends the well-known lemma of Makinson [11] for Boolean algebras to Heyting algebras. Corollary 3.4.19 is well known (see, e.g., Sec. III.3 of [1]).

3.5 Grzegorczyk Algebras

Grzegorczyk's modal system **Grz** is probably one of the most interesting systems (at least among the modal systems in Σ—see Definition 2.3.2). A. Grzegorczyk defined the modal system **Grz** axiomatically and showed that the system is a modal companion of intuitionistic logic, i.e., that for each formula p,

$$\mathbf{In} \vdash p \Leftrightarrow \mathbf{Grz} \vdash T(p);$$

here $T(p)$ is the modal formula obtained by the well-known Gödel translation of p.[19] It has been shown that the modal system **Grz** is the **strongest** modal companion of intuitionistic logic. This allows one to construct the following "superintuitionistic" classification of the family Σ (see Sect. 2.3) of all normal extensions of Lewis's classical modal system **S4**. For each superintuitionistic logic $\lambda \in \Lambda$, let Σ_λ denote the class of all **modal companions** of λ. That is, Σ_λ is the class of all modal systems $\sigma \in \Sigma$ such that the **superintuitionistic fragment** of σ coincides with λ.[20] The classes Σ_λ will be called **layers**. Clearly the classes Σ_λ ($\lambda \in \Lambda$) partition Σ into layers, which are in one-to-one correspondence with superintuitionistic logics. We point out the following extreme cases: [21]

(a) If $\lambda = \mathbf{CL}$ (classical two-valued logic), then the layer Σ_λ consists of all extensions of Lewis's system **S5**.[22]

(b) If $\lambda = \mathbf{In}$ (intuitionistic logic), then $\Sigma_\lambda = \{\sigma \in \Sigma : \mathbf{S4} \subseteq \sigma \subseteq \mathbf{Grz}\}$; so, intuitionistic logic can be embedded (by Gödel's translation) into any modal system situated between **S4** and **Grz**.

Furthermore, every layer Σ_λ ($\lambda \in \Lambda$) is an infinite closed interval (of a lattice Σ) of the form $[\mathbf{S4} + \Gamma, \mathbf{Grz} + \Gamma]$, i.e., $\Sigma_\lambda = \{\sigma \in \Sigma : \mathbf{S4} + \Gamma \subseteq \sigma \subseteq \mathbf{Grz} + \Gamma\}$, where Γ is a set of modal formulae. An important feature of Grzegorczyk's system is contained in the following statement: the partially ordered set $(\Lambda, \subseteq)$ is isomorphic to the partially ordered set of all normal extensions of Grzegorczyk's modal system **Grz**.[23]

3.5.1 Definition (*Grzegorczyk's modal system*) **Grz** is the normal extension of Lewis's system **S4** obtained by postulating the formula

$$(g_0) \qquad \Box(\Box(p \to \Box p) \to p) \to p$$

---

[19] *Editorial note*: The **Gödel translation** T is defined as follows: if p is a propositional variable, let $T(p) = \Box p$, and for arbitrary formulae p, q, let $T(p \vee q) = T(p) \vee T(q)$, $T(p \wedge q) = T(p) \wedge T(q)$, $T(p \to q) = \Box(T(p) \to T(q))$, and $T(\neg p) = \Box\neg T(p)$.

[20] *Editorial note*: The **superintuitionistic fragment** of σ is the superintuitionistic logic $\{p \mid T(p) \in \sigma\}$.

[21] *Editorial note*: For a proof, see Corollary 9.64 (and Exercise 9.21) in A. Chagrov and M. Zakharyaschev, *Modal Logic* (Clarendon Press), 1997.

[22] *Editorial note*: **S5** is obtained by adding to **S4** the axiom $\Diamond p \to \Box\Diamond p$.

[23] *Editorial note*: This result is now known as the Blok–Esakia Theorem; see p. 325 of A. Chagrov and M. Zakharyaschev, *Modal Logic* (Clarendon Press), 1997.

as a new axiom; symbolically, **Grz** $=$ **S4** $+ g_0$. More precisely, **Grz** is given by

$$\begin{array}{lll}\text{Axiom schemas:} & A1. & \Box(p \to q) \to (\Box p \to \Box q),\\ & A2. & \Box p \to p,\\ & A3. & \Box p \to \Box\Box p,\\ & A4. & \Box(\Box(p \to \Box p) \to p) \to p;\\ \text{Inference Rules:} & (R1) & \dfrac{p,\ p \to q}{q},\\ & (R2) & \dfrac{p}{\Box p}.\end{array}$$

3.5.2 Theorem *Grzegorczyk's modal system* **Grz** *can be defined axiomatically as follows:* **Grz** $=$ **S4** $+ (R3)$*, where*

$$(R3) \qquad \frac{\Box(p \to \Box p) \to p}{p}.$$

In other words, **Grz** *differs from* **S4** *only by the presence of the additional rule of inference* $(R3)$.

The proof of this theorem is based on the theory of hybrids and is deferred for a bit (see Lemma 3.5.11).

Note that Grzegorczyk's formula (g_0) $\Box(\Box(p \to \Box p) \to p) \to p$ can be rewritten in terms of the modal operator of "possibility" $\Diamond$ $(= \neg\Box\neg)$ as follows:

$$(g_1) \qquad p \to \Diamond(p \ \&\ \neg\Diamond(\neg p \ \&\ \Diamond p)).$$

Let $(B, \mathbf{C})$ be a closure algebra. The algebraic polynomial corresponding to Grzegorczyk's formula (g_0), as is easy to verify, looks as follows:

$$-a \vee \mathbf{C}(a \wedge -\mathbf{C}(-a \wedge \mathbf{C}a)) = 1,$$

or equivalently,

$$a \leq \mathbf{C}(a - \mathbf{C}(\mathbf{C}a - a)).$$

3.5.3 Definition The closure algebra $(B, \mathbf{C})$ is called a **Grzegorczyk algebra** if for each $a \in B$,

$$(g_2) \qquad a \leq \mathbf{C}(a - \mathbf{C}(\mathbf{C}a - a)).$$

It is clear from the definition that the class of Grzegorczyk algebras is a variety. We denote the class (and, as usual, the corresponding equational category) of Grzegorczyk algebras by GA. Clearly $\mathsf{GA} \subseteq \mathsf{CA}$ and the category GA of Grzegorczyk algebras is a full subcategory of the category CA of closure algebras.

Recall the notation of Sect. 2.5 (see Definition 2.5.12):

$$\rho a = a \wedge \mathbf{C}(\mathbf{C}a - a) \quad \text{and} \quad \pi a = a - \rho a.$$

In terms of the operator ρ, Grzegorczyk's "inequality" can be written as follows:

$$(g_3) \qquad a \leq \mathbf{C}\pi a.$$

Let $(B, \mathbf{C})$ be a Grzegorczyk algebra and (X, Ω, R) the corresponding hybrid. We use Convention 3.4.3 to identify $(B, \mathbf{C})$ with the algebra of clopen sets $(B(X), R^{-1})$ of the hybrid (X, Ω, R). Let A be a clopen set in (X, Ω, R), i.e., $A \in B(X) = B$. Then $\pi A = A - R^{-1}(R^{-1}A - A)$. In other words, for each $x \in X$,

$$x \in \pi A \Leftrightarrow (x \in A \ \& \ \neg(\exists y)(xRy \ \& \ y \notin A \ \& \ (\exists z)(yRz \ \& \ z \in A)).$$

3.5.4 Definition If A is clopen in X, then the points in πA are called **passive points** of A, and the points in ρA are called **active points** of A.

This can be explained as follows. A point $x \in A$ is a passive point of A (i.e., $x \in \pi A$) if and only if it is impossible, starting from x (see the right-hand side of the above equivalence) to "leave" the set A and then "return" to A again. On the other hand, a point $x \in A$ is an active point of A if it is possible starting from x to leave A and return to A again. Indeed,

$$x \in \rho A \Leftrightarrow (x \in A \ \& \ (\exists y \in X)(xRy \ \& \ y \notin A \ \& \ (\exists z \in X)(yRz \ \& \ z \in A)).$$

In terms of passive points, axiom (g_3) can be written as follows: starting from each point of a clopen set A, a passive point of A is reachable; more formally,

$$(g_3') \qquad (\forall x \in X)(x \in A \Rightarrow (\exists y \in X)(xRy \ \& \ y \in \pi A)).$$

Consequently, we established the following.

3.5.5 Theorem *Let $(B, \mathbf{C})$ be a closure algebra and (X, Ω, R) the corresponding hybrid. Then $(B, \mathbf{C})$ is a Grzegorczyk algebra if and only if for each clopen set A in (X, Ω, R), we have $A \subseteq R^{-1}(\pi A)$ (i.e., the set of all passive points of A is cofinal in A).*

The following theorem is as convenient.

3.5.6 Theorem *Let $(B, \mathbf{C})$ be a closure algebra and (X, Ω, R) the corresponding hybrid. The following are equivalent:*

(1) *Every quasi-maximal point of each clopen A is maximal;*
(2) *$(B, \mathbf{C})$ is a Grzegorczyk algebra.*

Proof $(1 \Rightarrow 2)$ Suppose (1) holds. Let us prove (g_3'). Let $x \in A$. By the remark after Theorem 3.2.1 and Corollary 3.2.2, there is a quasi-maximal point $y \in A$ such that

xRy. By (1), y is maximal. Clearly one cannot leave A and return to A starting from y, for this would contradict the maximality of y in A.

($2 \Rightarrow 1$) Suppose (1) is false, i.e., some clopen set A contains a quasi-maximal point $x \in A$ that is not maximal. This means that there exists $y \in X$ such that $x \neq y$, xRy, and yRx. Since $x \neq y$, there exists a clopen set A_0 such that $x \in A_0$ and $y \notin A_0$. Let us show that (g_3') is false for $x \in A \cap A_0$. Indeed, $x \in A \cap A_0$, and starting at x, one can leave $A \cap A_0$ (i.e., xRy and $y \notin A \cap A_0$) and return to $A \cap A_0$ again (yRx and $x \in A \cap A_0$). So, $(B, \mathbf{C})$ is not a Grzegorczyk algebra. □

3.5.7 Corollary *Let $(B, \mathbf{C}) \in \mathsf{CA}$ and (X, Ω, R) be the corresponding hybrid. Then $(B, \mathbf{C})$ is a Grzegorczyk algebra if and only if either of the following holds:*

(1) *for each clopen subset A of X, we have $A \subseteq R^{-1}(\max A)$;*
(2) *for each nonempty clopen set A, we have $\max A \neq \varnothing$.*

A close link between Grzegorczyk algebras and Heyting algebras is established by the following.

3.5.8 Corollary *If a hybrid (X, Ω, R) is strict, then the corresponding algebra $(B, \mathbf{C})$ is a Grzegorczyk algebra.*

3.5.9 Corollary *All skeletal closure algebras belong to the variety of Grzegorczyk algebras, i.e., $\mathsf{SA} \subseteq \mathsf{GA}$.*

Recall that a closure algebra $(B, \mathbf{C})$ is finite if and only if the corresponding hybrid (X, Ω, R) is a finite quasi-ordered set (X, R) with the discrete topology Ω. In this case we forget Ω (see Theorem 3.4.17(2)).

3.5.10 Corollary *Let $(B, \mathbf{C})$ be a finite closure algebra and (X, R) the corresponding quasi-ordered set. The following are equivalent:*

(1) *$(B, \mathbf{C})$ is a Grzegorczyk algebra;*
(2) *$(B, \mathbf{C})$ is a skeletal algebra;*
(3) *(X, R) is a finite partially ordered set.*

The following lemma is an algebraic formulation of Theorem 3.5.2.[24]

3.5.11 Lemma *Let $(B, \mathbf{C})$ be a closure algebra. The following are equivalent:*

(1) *$(B, \mathbf{C})$ is a Grzegorczyk algebra;*
(2) *for each $a \in B$, if $a \neq 0$, then $\pi a \neq 0$;*
(3) *for each $a \in B$, if $a \neq 0$, then $\rho a \neq a$;*
(4) *there are no cyclic elements in $(B, \mathbf{C})$.*

[24] *Editorial note*: Lemma 3.5.11 is an algebraic formulation of Theorem 3.5.2 provided that the rule $(R3)$ is reformulated in terms of $\Diamond$.

Proof (1 ⇒ 2) Let $(B, \mathbf{C})$ be a Grzegorczyk algebra, i.e., assume that for all $a \in B$, we have $a \leq \mathbf{C}(a - (\mathbf{C}a - a))$. Let $a \neq 0$. Suppose $\pi a = a - \mathbf{C}(\mathbf{C}a - a) = 0$. Then $\mathbf{C}\pi a = 0$ and $0 \neq a \leq \mathbf{C}\pi a = 0$, a contradiction. Hence, $\pi a \neq 0$.

(2 ⇒ 1) Identify the algebra $(B, \mathbf{C})$ with the algebra of clopen sets of its hybrid (X, Ω, R). Suppose (1) fails, i.e., there exists a clopen set A such that $A \not\subseteq R^{-1}(\pi A)$, so $A - R^{-1}(\pi A) \neq \varnothing$. Then $A \neq \varnothing$. Let $A_0 = A - R^{-1}(\pi A)$. Since A is clopen and R^{-1} maps clopen sets to clopen sets (Theorem 3.2.1(IIIb)), we see that both πA and πA_0 are clopen. Now suppose $\pi A_0 \neq \varnothing$. Then there is $x \in \pi A_0$, i.e., x is a passive point of $A_0 = A - R^{-1}(\pi A)$. We have $x \in A$ and $R(x) \cap \pi A = \varnothing$. Therefore, $x \notin \pi A$. But then x is an active point of A, so there must be a point $y \in X$ such that xRy, $y \notin A$, and $(\exists z)(yRz \;\&\; z \in A)$. Since R is transitive, we obtain xRz and $z \in A$. As $R(x) \cap \pi A = \varnothing$, we have $z \notin \pi A$. Moreover, $z \notin R^{-1}(\pi A)$, for otherwise, $R(z) \cap \pi A \neq \varnothing$ and (since xRz) $R(x) \cap \pi A \neq \varnothing$. Furthermore, since $z \notin R^{-1}(\pi A)$ and $z \in A$, we have $z \in A_0$. So xRy and $y \notin A$, and hence $y \notin A_0$, as well as yRz and $z \in A_0$. Therefore, starting at x, one can leave A_0 (xRy and $y \notin A_0$) and return to A_0 again (yRz and $z \in A_0$). It follows that x is an active (and not passive) point of A_0, so $x \notin \pi A_0$, a contradiction. Thus, $\pi A_0 = \varnothing$, and we have found a nonempty clopen A_0 such that $\pi A_0 = \varnothing$. This disproves (2). By contraposition, (2) implies (1).

(2 ⇔ 3) Let (2) hold and $a \neq 0$. Suppose $\rho a = a$. Then $\pi a = 0$ since $\pi a = a - \rho a$, a contradiction. Conversely, let (3) hold and $a \neq 0$. Then $\pi a = a - \rho a \neq 0$ because $\rho a \leq a$ and $\rho a \neq a$. Thus, $\pi a \neq 0$.

(3 ⇔ 4) Suppose (3) holds and there exists a cyclic element $b \in B$; that is, $\rho^m(b) = \rho^{m+1}(b) \neq 0$. Then $a = \rho^m(b) \neq 0$ and $\rho a = a$, contradicting (3). Conversely, suppose (3) fails, i.e. there exists $a \neq 0$ such that $\rho a = a$. Clearly a is cyclic. □

Grzegorczyk algebras are then characterized by the following.

3.5.12 Corollary *A closure algebra $(B, \mathbf{C})$ is a Grzegorczyk algebra if and only if it has no cyclic elements.*

3.5.13 Theorem *The variety* GA *of Grzegorczyk algebras is generated by its finite algebras.*[25]

Proof First note that if p, q are algebraic polynomials of a closure algebra, then $p = q$ is equivalent to $(-p \vee q) \wedge (-q \vee p) = 1$. Indeed, $p = q \Leftrightarrow p \leq q$ and $q \leq p \Leftrightarrow (-p \vee q) = 1$ and $(-q \vee p) = 1 \Leftrightarrow (-p \vee q) \wedge (-q \vee p) = 1$.

Now let p be a polynomial. If $p = 1$ in all Grzegorczyk algebras, then clearly $p = 1$ in all finite Grzegorczyk algebras. Now let $p = p(v_1, \ldots, v_k) \neq 1$ (where $v_1, \ldots, v_k$ are the variables of p) in some algebra $(B, \mathbf{C}) \in$ GA. Then there exist $a_1, \ldots, a_k \in B$ such that $p(a_1, \ldots, a_k) \neq 1$. Identify the algebra $(B, \mathbf{C})$ with the

[25] *Editorial note*: Julia Ilin pointed out a gap in Esakia's original proof, so we have replaced it by an algebraic version of the proof from pp. 158–9 of G. Boolos, *The Logic of Provability* (Cambridge University Press), 1993.

algebra of clopen sets $(B(X), R^{-1})$ of its hybrid (X, Ω, R). Let S be the set of subpolynomials of $p(a_1, \ldots, a_k)$ and

$$T = S \cup \{\mathbf{C}(\mathbf{C}q - q) : q \in S\}.$$

Let B_0 be the Boolean subalgebra of B generated by T. Since T is finite, B_0 is finite, and it induces an equivalence relation $\sim$ on X given by

$$x \sim y \text{ iff } (\forall a \in B_0)(x \in a \Leftrightarrow y \in a).$$

Let $[x]$ be the equivalence class of x, and $X_0 = \{[x] \mid x \in X\}$. The equivalence classes are exactly the atoms of B_0, so every element of B_0 is a finite union of equivalence classes. Thus, B_0 is isomorphic to $\mathrm{Sub}(X_0)$.

Define a relation R_0 on X_0 by:

$$\begin{aligned}[x]R_0[y] \quad &\text{iff} \quad (\forall a)[(\mathbf{C}a \in B_0 \ \& \ a \in y) \Rightarrow \mathbf{C}a \in x] \\ &\text{iff} \quad (\forall a)[(\mathbf{C}a \in B_0 \ \& \ \mathbf{C}a \in y) \Rightarrow \mathbf{C}a \in x],\end{aligned}$$

where the equivalence of the two definitions follows from the axioms $a \leq \mathbf{C}a$ and $\mathbf{CC}a \leq \mathbf{C}a$, which also imply that R_0 is reflexive and transitive. However, R_0 is not in general a partial order, so we define

$$[x]R_1[y] \quad \text{iff} \quad [x]R_0[y] \text{ and } ([y]R_0[x] \Rightarrow [x] = [y]).$$

We claim that R_1 is a partial order. Reflexivity is obvious and antisymmetry is immediate from the definition. For transitivity, suppose $[x]R_1[y]R_1[z]$, which implies $[x]R_0[y]R_0[z]$ and hence $[x]R_0[z]$. If in addition $[z]R_0[x]$, then $[z]R_0[y]$ by the transitivity of R_0, which with $[y]R_1[z]$ implies $[z] = [y]$. Then from $[x]R_1[y]$ we have $[x]R_1[z]$.

Define $\mathbf{C}_1$ on B_0 by

$$\mathbf{C}_1 a = \bigcup \{[x] : \exists y \in a \text{ such that } [x]R_1[y]\}.$$

Since B_0 is isomorphic to $\mathrm{Sub}(X_0)$, it follows from the definition of $\mathbf{C}_1$ that $(B_0, \mathbf{C}_1)$ is isomorphic to $(\mathrm{Sub}(X_0), R_1^{-1})$. Then since (X_0, R_1) is a finite partially ordered set, $(B_0, \mathbf{C}_1)$ is a finite Grzegorczyk algebra by Corollary 3.5.10.

We claim that for each $\mathbf{C}q \in S$, we have $\mathbf{C}q = \mathbf{C}_1 q$. First, we show that $x \notin \mathbf{C}q$ implies $x \notin \mathbf{C}_1 q$. Suppose $x \notin \mathbf{C}q$. To show $x \notin \mathbf{C}_1 q$, we must show that for any y with $[x]R_1[y]$, we have $y \notin q$. So suppose $[x]R_1[y]$ and hence $[x]R_0[y]$. If $y \in q$, then since $[x]R_0[y]$ and $\mathbf{C}q \in B_0$, we have $x \in \mathbf{C}q$, contradicting our assumption. Thus, $y \notin q$, so we conclude that $x \notin \mathbf{C}_1 q$.

Second, we show that $x \in \mathbf{C}q$ implies $x \in \mathbf{C}_1 q$. Assume $x \in \mathbf{C}q$. If $x \in q$, we are done, so assume $x \notin q$. Thus, $x \in \mathbf{C}q - q$ and hence $x \in \mathbf{C}(\mathbf{C}q - q)$. Since $x \in \mathbf{C}q$, there is a $y \in q$ such that xRy. Then by (g_2), $y \in \mathbf{C}(q - \mathbf{C}(\mathbf{C}q - q))$, so there is a $z \in q - \mathbf{C}(\mathbf{C}q - q)$ such that yRz. Hence xRz, which implies $[x]R_0[z]$. Since $\mathbf{C}q \in S$ and hence $q \in S$, we have $\mathbf{C}(\mathbf{C}q - q) \in T$. Then given $\mathbf{C}(\mathbf{C}q - q) \in B_0$, $x \in \mathbf{C}(\mathbf{C}q - q)$, and $z \notin \mathbf{C}(\mathbf{C}q - q)$, it is not the case that $[z]R_0[x]$. Thus $[x]R_1[z]$, which with $z \in q$ implies $x \in \mathbf{C}_1 q$.

As a consequence of the previous claim and the fact that $p(a_1, \ldots, a_k) \neq 1$ in $(B, \mathbf{C})$, we conclude that $p(a_1, \ldots, a_k) \neq 1$ in $(B_0, \mathbf{C}_1)$. Since $(B_0, \mathbf{C}_1)$ is a finite Grzegorczyk algebra, this completes the proof. □

3.5.14 Corollary *The variety* GA *of Grzegorczyk algebras is generated by the class of all skeletal algebras.*

Proof By Corollary 3.5.9, SA $\subseteq$ GA. By Corollary 3.5.10, every finite Grzegorczyk algebra is a skeletal algebra. Thus, the result follows from Theorem 3.5.13. □

We now give a name to the hybrids dual to Grzegorczyk algebras.

3.5.15 Definition A hybrid (X, Ω, R) is **well founded** if, starting at any point x of any clopen set A, one can "reach" a maximal point of A; more precisely, if $A \subseteq R^{-1}(\max A)$.

Let HYB^f be the category of all well founded hybrids and hybrid maps.

3.5.16 Theorem *The category* GA *of Grzegorczyk algebras is dual to the category* HYB^f *of well founded hybrids.*

Proof Apply the Duality Theorem (Theorem 3.3.4), Theorem 3.5.6, and Corollary 3.5.7. □

Using this theorem and Corollary 3.5.8, we easily obtain the following.

3.5.17 Corollary *The category* HYB^+ *of strict hybrids is a full subcategory of the category* HYB^f *of well founded hybrids.*

Bibliographic notes. Grzegorczyk algebras were introduced by the author as algebraic models for one of the most interesting modal systems (at least among the systems of the post-Lewis period), namely the system formulated axiomatically by A. Grzegorczyk in [8]. Some results were reported by the author in [7]; see also [6]. Segerberg [17] was the first to prove the finite approximability of Grzegorczyk's modal system in terms of the semantics of "possible worlds" (cf. Theorem 3.5.13).[26] A number of deep results related to Grzegorczyk algebras appeared in Blok's monograph [2]. Some results obtained by L. Maksimova in the interesting papers [12, 13] are related to Grzegorczyk algebras.

[26] *Editorial note*: A logic is **finitely approximable** if it is characterized (in the sense of Sect. 2.3.3) by a class of finite algebras.

References

1. Birkhoff, G. (1967). *Lattice theory* (Vol. 3), American mathematical society colloquium publications Providence: American Mathematical Society.
2. Blok, W.J. (1976) *Varieties of interior algebras*. Ph.D. thesis, University of Amsterdam, Amsterdam.
3. Cohn, P. M. (1965). *Universal algebra*. New York: Harper & Row Publishers.
4. Esakia, L. L. (1974). Topological Kripke models. *Soviet Mathematics Doklady*, *15*, 147–151.
5. Esakia, L. L. (1979). On the theory of modal and superintuitionistic systems (Russian). *Logical inference* (pp. 147–172). Moscow: Nauka.
6. Esakia, L. L. (1981). Diagonal constructions, Löb's formula, and Cantor's scattered spaces (Russian). *Studies in logic and semantics* (pp. 128–143). Tbilis: Metsniereba.
7. Esakia, L. L. (1984). On the variety of Grzegorczyk algebras. *Selecta Sovietica Mathematica*, *3*, 343–366.
8. Grzegorczyk, A. (1967). Some relational systems and the associated topological spaces. *Fundamenta Mathematicae*, *60*, 223–231.
9. Halmos, P. R. (1962). *Algebraic logic*. New York: Chelsea Publishing Co.
10. Jónsson, B., & Tarski, A. (1951). Boolean algebras with operators. I. *American Journal of Mathematics*, *73*, 891–939.
11. Makinson, D. C. (1969). On the number of ultrafilters of an infinite Boolean algebra. *Zeitschrift für Mathematische Logik und Grundlagen der Mathematik*, *15*, 121–122.
12. Maksimova, L. L. (1979). Interpolation theorems in modal logics and amalgamable varieties of topological Boolean algebras. *Algebra and Logic*, *18*, 348–370.
13. Maksimova, L. L., & Rybakov, V. V. (1974). The lattice of normal modal logics. *Algebra and Logic*, *13*, 105–122.
14. Mostowski, A., & Tarski, A. (1939). Boolesche ringe mit geordneter basis. *Fundamenta Mathematicae*, *32*, 69–86.
15. Nachbin, L. (1965) *Topology and order*. Translated from the Portuguese by Lulu Bechtolsheim. Van Nostrand Mathematical Studies (Vol. 4). Princeton: D. Van Nostrand Co., Inc.
16. Priestley, H. A. (1972). Ordered topological spaces and the representation of distributive lattices. *Proceedings of the London Mathematical Society*, *3*(24), 507–530.
17. Segerberg, K. (1971) *An essay in classical modal logic. Vols. 1, 2, 3*. Filosofiska Studier (Vol. 13). Uppsala: Filosofiska Föreningen och Filosofiska Institutionen vid Uppsala Universitet.
18. Sikorski, R. (1960). *Boolean algebras* (Vol. 25). Ergebnisse der Mathematik und ihrer Grenzgebiete, N. F. Berlin: Springer.
19. Wolk, E. S. (1968). The topology of a partially well ordered set. *Fundamenta Mathematicae*, *62*, 255–264.

Appendix

In this appendix, we outline, in general terms, the content of the second part of the monograph entitled 'Heyting Algebras II. Additional Chapters' which is being prepared for publication.[1] Hopefully, the appendix will allow the reader to form an opinion about the topics covered in part II.

A.1 On Subdirectly Irreducible Heyting Algebras and Subdirect Products

In Sect. 1.1, we defined subdirect products of algebras and subdirectly irreducible algebras, and stated a number of known theorems, including G. Birkhoff's classical theorem. It was noted that every variety is generated by its subdirectly irreducible algebras; and, in a sense, every variety is characterized completely by its subdirectly irreducible members.

Let us give a characterization of subdirectly irreducible Heyting algebras and closure algebras in terms of hybrids.

A closure algebra $(B, \mathbf{C})$ (resp. a Heyting algebra H) is called **well-connected** [14] if for any $a, b \in B$ (resp. $a, b \in H$) $\mathbf{C}a \wedge \mathbf{C}b = 0$ (resp. $a \vee b = 1$) implies that either $\mathbf{C}a = 0$ or $\mathbf{C}b = 0$ (resp. either $a = 1$ or $b = 1$).

It is easy to see that each subdirectly irreducible Heyting algebra (see Theorem 2.4.11) is well-connected; hence, every variety of Heyting algebras is generated by its well-connected algebras. In light of Corollary 2.4.21, every subdirectly irreducible closure algebra $(B, \mathbf{C})$ is well-connected.[2] A characterization of subdirectly irreducible and well-connected algebras is given by the following.

[1] *Editorial note*: As noted in the Foreword, volume II was never published.

[2] *Editorial note*: In addition to Corollary 2.4.21, this uses the equivalence of (1) and (2) in Proposition A.1.1.

G. Bezhanishvili and W. H. Holliday (eds.), *Heyting Algebras*,
Trends in Logic 50, https://doi.org/10.1007/978-3-030-12096-2

A.1.1 Proposition *Let $(B, \mathbf{C})$ be a closure algebra and H its skeleton. Then the following are equivalent:*

(1) *H is well-connected;*
(2) *$(B, \mathbf{C})$ is well-connected;*
(3) *the hybrid (X, Ω, R) that corresponds to the algebra $(B, \mathbf{C})$ is a principal cone, i.e., there exists $x \in X$ such that $R(x) = X$.*

A.1.2 Proposition (Subdirect irreducibility) *Let $(B, \mathbf{C})$ be a closure algebra and H its skeleton. Then the following are equivalent:*

(1) *H is subdirectly irreducible;*
(2) *$(B, \mathbf{C})$ is subdirectly irreducible;*
(3) *the hybrid (X, Ω, R) is* **super-principal**, *i.e., there exists a clopen cluster C such that $R(x) = X$ for each $x \in C$.*

Using Theorem 3.4.16 (see also Definition 3.4.12), one can prove:

A.1.3 Proposition (Subdirect products) *The following are equivalent:*

(1) *the Heyting algebra H is decomposable into a subdirect product of algebras H_i $(i \in I)$;*
(2) *let A_i be the closed cone of the hybrid (X, Ω, R) corresponding to the Heyting algebra H_i $(i \in I)$; then the family of cones A_i is a covering of some dense subspace X_0 of the space (X, Ω), i.e., $X_0 \subseteq \bigcup\{A_i : i \in I\}$.*

A.1.4 Proposition (Cone covering) *A Heyting algebra H is finitely approximable if and only if the corresponding hybrid (X, Ω, R) contains a family $\{A_i : i \in I\}$ of finite cones covering some dense subset X_0 of the space X, i.e., $X_0 \subseteq \bigcup\{A_i : i \in I\}$.*

A.2 Hybrid Formulation of Glivenko's Theorem

Let H be a Heyting algebra and $Rg(H)$ the set of all its **regular elements**, i.e.,

$$Rg(H) = \{\neg a : a \in H\}.$$

It is known that $Rg(H)$ is a Boolean algebra and that the map $h : H \to Rg(H)$, given by $ha = \neg\neg a$ $(a \in H)$, is a surjective homomorphism of the Heyting algebra H onto the Boolean algebra $Rg(H)$ (see [16]).

A.2.1 Proposition *Let (X, Ω, R) be the hybrid corresponding to the Heyting algebra H. Then the set $Y = \max X$ is a Stone space and the Boolean algebra $B(Y)$ of clopen sets of Y is isomorphic to the algebra $Rg(H)$. Note that the hybrid (Y, Ω_Y, R_Y) is a subhybrid of (X, Ω, R), where $Y = \max X$, Ω_Y is the subspace topology, and R_Y is the restriction of R to Y. Moreover, R_Y is the identity relation on Y.*

A.3 Weak Decompositions of Boolean Algebras and Heyting Algebras

The content of this section is based on [10]. Let $\{H_i : i \in I\}$ be a family of Heyting algebras. For each $i \in I$, let p_i be the projection of the direct product $\prod_{i \in I} H_i$ onto H_i.

A subalgebra H of the direct product $\prod_{i \in I} H_i$ is called a **weak product** of the Heyting algebras H_i $(i \in I)$ if H satisfies the following **absorption** condition: if $a \in \prod_{i \in I} H_i$ differs from $b \in H$ only by finitely many components, then a is "absorbed" by H. More precisely, if $a \in \prod_{i \in I} H_i$, $b \in H$ and the set $\{i \in I : p_i(a) \neq p_i(b)\}$ is finite, then $a \in H$.

Note that the concept of the weak product is applicable to bounded distributive lattices as well. It is easy to check that every weak product of Heyting algebras H_i $(i \in I)$ is a subdirect product of these algebras. If I is finite, then the direct product of the algebras H_i $(i \in I)$ is their only weak product. Let $\prod^W_{i \in I} H_i$ denote an arbitrary weak product of the algebras H_i.

A set $S = \{a_i \in H : i \in I\}$ of non-zero elements of H is called a **partition of the unit** if $a_i \wedge a_j = 0$ for all $i \neq j$ and $\bigvee\{a_i : i \in I\} = 1 \in H$.

An internal characterization of weak decompositions of Heyting algebras is given by the following statement.

A.3.1 Proposition (Weak decompositions) *A Heyting algebra H is isomorphic to a weak product $\prod^W_{i \in I} H_i$ of Heyting algebras H_i $(i \in I)$ if and only if there exists a partition of the unit $\{a_i : i \in I\}$ in H such that $H/a_i \simeq H_i$ for all $i \in I$ (where H/a_i is the quotient algebra of H by the principal filter $[a_i)$).*

The following corollary is of independent interest.

A.3.2 Corollary (Weak decompositions for Boolean algebras) *A Boolean algebra B is isomorphic to a weak product $\prod^W_{i \in I} B_i$ of Boolean algebras B_i $(i \in I)$ if and only if there exists a partition of the unit $\{a_i : i \in I\}$ of B such that $B/a_i \simeq B_i$ for all $i \in I$.*

Let $\bigoplus_{i \in I} H_i$ be the **direct sum** of Heyting algebras H_i $(i \in I)$, i.e., the subalgebra of their direct product $\prod_{i \in I} H_i$ consisting of all elements $a \in \prod_{i \in I} H_i$ such that at least one of $\{i \in I : p_i(a) \neq 0\}$, $\{i \in I : p_i(a) \neq 1\}$ is finite. It is easy to check that the direct sum $\bigoplus_{i \in I} H_i$ is a weak product of the algebras H_i $(i \in I)$.

A.3.3 Proposition *Weak products of the algebras H_i $(i \in I)$ are precisely the subalgebras of their direct product $\prod_{i \in I} H_i$ that contain their direct sum $\bigoplus_{i \in I} H_i$ as a subalgebra. In other words, weak products of Heyting algebras H_i $(i \in I)$ are the* *algebras H "situated between" the direct sum $\bigoplus_{i \in I} H_i$ and the direct product $\prod_{i \in I} H_i$, i.e., $\bigoplus_{i \in I} H_i \subseteq H \subseteq \prod_{i \in I} H_i$.*

One can describe weak products in terms of duality theory as follows. For simplicity and brevity, we restrict ourselves to Boolean algebras. Let B_i $(i \in I)$ be a Boolean algebra and X_i the corresponding Stone space. Let $\sum_{i \in I} X_i$ be the topological sum of the spaces X_i. It is known that $\sum_{i \in I} X_i$ is zero-dimensional and Hausdorff. Note that if the set of indices I is infinite, then $\sum_{i \in I} X_i$ is not compact. Let $\beta\big(\sum_{i \in I} X_i\big)$ be the Stone–Čech compactification[3] of $\sum_{i \in I} X_i$. For a space X, let $\mathcal{Z}(X)$ be the set of all zero-dimensional Hausdorff compactifications[4] of X. The set $\mathcal{Z}(X)$ can be partially ordered by means of continuous maps identical on X. Finally, let W be the set of all weak products of the family $\{B_i : i \in I\}$ of Boolean algebras ordered by inclusion.

A.3.4 Proposition *The partially ordered set W is isomorphic to the partially ordered set $\mathcal{Z}\big(\sum_{i \in I} X_i\big)$ of compactifications of the topological sum $\sum_{i \in I} X_i$ of Stone spaces X_i.*

A.3.5 Corollary (Ph. Dwinger) *The Stone space X of the direct product $\prod_{i \in I} B_i$ of Boolean algebras B_i $(i \in I)$ is homeomorphic to the Stone–Čech compactification $\beta\big(\sum_{i \in I} X_i\big)$ of the topological sum of the spaces X_i, where X_i corresponds to the Boolean algebra B_i.*[5]

A.3.6 Corollary (G. Day) *The Stone space X of the direct sum $\bigoplus_{i \in I} B_i$ of Boolean algebras B_i $(i \in I)$ is homeomorphic to the Alexandroff one-point compactification $\alpha\big(\sum_{i \in I} X_i\big)$ of the topological sum $\sum_{i \in I} X_i$.*[6]

Concluding this section, we note that if a Heyting algebra H is the direct product $\prod_{i \in I} H_i$ of algebras H_i $(i \in I)$, then the skeletal closure algebra $B(H)$ is isomorphic to a weak product $\prod_{i \in I}^{W} B(H_i)$ of skeletal algebras $B(H_i)$ (cf. Proposition 2.5.9(2)). This weak product coincides with the *categorical* product of the algebras $B(H_i)$ in the category SA (and in general does *not* coincide with their direct product; recall that the class SA of skeletal algebras is *not* a variety). Using the above, one can give a description of the direct product of Heyting algebras in terms of hybrids.

---

[3] *Editorial note*: See, e.g., Sect. 3.6 of R. Engelking, *General Topology*, 2nd ed. (Heldermann-Verlag), 1989.

[4] *Editorial note*: For the notion of compactification, see, e.g., Sect. 3.5 of R. Engelking, *General Topology*, 2nd ed. (Heldermann-Verlag), 1989.

[5] *Editorial note*: See Theorem 3 of Ph. Dwinger, "Remarks on the field representations of Boolean algebras," *Indagationes Mathematicae* 22: 213–217, 1960.

[6] *Editorial note*: See Theorem 4 of G. Day, "Superatomic Boolean algebras," *Pacific Journal of Mathematics* 23(3): 479–489, 1967.

A.4 The Lattice of Varieties of Heyting Algebras

Using the descriptions of weak products and almost direct products (in the sense of S. Feferman and R. Vaught)[7] of closure algebras and Heyting algebras in terms of hybrids [7], one can prove the following statement, due to W. Blok [3] and the author [4].[8]

A.4.1 Proposition *The lattice* $L(\mathsf{HA})$ *of varieties of Heyting algebras is isomorphic to the lattice* $L(\mathsf{GA})$ *of varieties of Grzegorczyk algebras.*

A.5 Representable Ordered Sets: Grätzer's Problems

A partially ordered set (X, R) is said to be **hybrid** (or **representable**) if there exists a hybrid (X', Ω', R') such that the partially ordered set (X, R) is isomorphic to (X', R'). Let HPO be the class of all hybrid partially ordered sets. Clearly a partially ordered set (X, R) is hybrid if and only if there exists a Heyting algebra H such that (X, R) is isomorphic to the partially ordered set of all prime filters of H.

A.5.1 Proposition *Let* (X, R) *be a hybrid partially ordered set (i.e.,* $(X, R) \in \mathsf{HPO}$*).*

(1) *The partially ordered set* (X, R) *has enough gaps.*[9]
(2) *The partially ordered set* (X, R) *is Dedekind complete.*[10]
(3) *All finite partially ordered sets belong to* HPO.
(4) *A linearly ordered set* (X, R) *belongs to* HPO *if and only if* (X, R) *satisfies conditions* (1)–(2).[11]
(5) *The class* HPO *is closed under direct products of arbitrary families of partially ordered sets from* HPO.
(6) *The class* HPO *is closed under cardinal sums of arbitrary families of partially ordered sets from* HPO *(recall that* (X, R) *is the* **cardinal sum** *of partially ordered sets* (X_i, R_i) $(i \in I)$ *if* X *is the (disjoint) union of the sets* X_i $(i \in I)$ *and for all* $x, y \in X$, $xRy \Leftrightarrow \exists i \in I : x, y \in X_i \ \& \ xR_iy$*).*

Unfortunately, the author failed to obtain a "digestable" characterization of the class HPO.

We conclude the section by quoting Grätzer [12]: "Investigate further the poset $\mathscr{P}(L)$ of prime ideals of a distributive lattice" (Problem II.4) and "Characterize

[7] *Editorial note*: See S. Feferman and R. L. Vaught, "The first order properties of products of algebraic systems," *Fundamenta Mathematicae* 47: 57–103, 1959.

[9] *Editorial note*: Recall Definition 3.2.13.

[10] *Editorial note*: Recall Definition 3.2.18.

[11] *Editorial note*: C.f. Theorem 9 of R. Balbes, "On the Partially Ordered Set of Prime Ideals of a Distributive Lattice," *Canadian Journal of Mathematics* 23(5): 866–874, 1971.

$\mathscr{P}(L)$ under the additional assumption that L has a unit and/or a zero" (Problem II.5). It is tempting to replace L by H and $\mathscr{P}(L)$ by $\mathscr{P}(H)$ in those quotations and suggest this as a new problem to the reader.

A.6 MacNeille Completions

MacNeille showed, by generalizing the celebrated Dedekind cuts, that every partially ordered set can be embedded into a complete lattice so that all existing suprema and infima are preserved (the so-called **MacNeille completion**; see for example Sect. V.9 of [2]). However, this embedding does not preserve certain important properties: e.g., there is an example of a distributive lattice whose MacNeille completion is not even modular (N. Funayama). The MacNeille completion of a Boolean lattice is a Boolean lattice (V. Glivenko, M. Stone). Using the Duality Theorem and certain b properties of hybrids, one can describe a construction dual to MacNeille's construction and show that the completion of a Heyting lattice is a Heyting lattice.

A.7 The Category of Heyting Algebras is Balanced

The content of this section is based on [11]. Let (X_i, Ω_i, R_i) be hybrids $(i = 1, 2)$. A hybrid (X, Ω, R) is called the **product** of the hybrids (X_i, Ω_i, R_i) $(i = 1, 2)$ if the set $X = X_1 \times X_2$ is equipped with the topology Ω of the product of the spaces (X_i, Ω_i) $(i = 1, 2)$ and the relation R on X is defined componentwise, i.e.,

$$(x, y)R(x', y') \Leftrightarrow xR_1x' \;\;\&\;\; yR_2y'.$$

Let X_0, X_1, X_2 be hybrids, $f_i : X_i \to X_0$ a surjective hybrid map $(i = 1, 2)$, and $X = \{(x, y) \in X_1 \times X_2 : \; f_1(x) = f_2(y)\}$.

A.7.1 Proposition *Let H_0, H_1, H_2 be the Heyting algebras corresponding to the hybrids X_0, X_1, and X_2, and let $h_i : H_0 \to H_i$ $(i = 1, 2)$ be the inclusions dual to the maps f_i. If H is the Heyting algebra of cones of X, then H amalgamates the amalgam (H_0, H_1, H_2).*

Using the Duality Theorem (Theorem 3.3.4), Proposition A.7.1, and some properties of hybrids, we can prove:

A.7.2 Proposition *The category* HA *of Heyting algebras is balanced; that is, every epimorphism of the category* HA *is surjective.*

We recall that the category BA of Boolean algebras is balanced, but that the category DL of bounded distributive lattices is not balanced.

A.8 Equationally Non-conservative Operations

It is well known that every equational class (variety) K of algebras is closed under the fundamental algebraic operations of taking homomorphic images (**H**), subalgebras (**S**), direct products (**P**), subdirect products ($\mathbf{P_S}$), direct limits $\left(\underrightarrow{\lim}\right)$, and inverse limits $\left(\underleftarrow{\lim}\right)$. In order to emphasize this conservation property, i.e., the preservation of identities under the indicated operations, we call such operations **equationally conservative** (or simply **conservative**). The role of such operations in modern algebraic research is well known. However, we can mention interesting applications of **non-conservative** operations as well.

(A) Tensor sums [8]. Let $\{X_i : i \in I\}$ be a family of hybrids ($I \neq \varnothing$) and $X = \prod\limits_{i \in I} X_i$ the cartesian product of the sets X_i $(i \in I)$. Equip X with the product topology Ω and the "componentwise" order relation R, i.e., $xRy \Leftrightarrow x_i R_i y_i$ for all $i \in I$. Call X the **product** of the hybrids X_i $(i \in I)$.

A.8.1 Proposition (a) *If X_i is a hybrid for each $i \in I$, then the product $X = \prod\limits_{i \in I} X_i$ is a hybrid;*
(b) *the hybrid X is strict if and only if X_i is strict for each $i \in I$.*

Let B_i be the closure algebra (resp. let H_i be the Heyting algebra) corresponding to the hybrid X_i $(i \in I)$. The closure algebra B (resp. the Heyting algebra H) dual to the hybrid (resp. the strict hybrid) X is called the **tensor sum** of closure algebras B_i (resp. the Heyting algebras H_i) and is denoted by $B = \bigotimes\limits_{i \in I} B_i$ (resp. $H = \bigotimes\limits_{i \in I} H_i$).

A.8.2 Proposition *The variety of Heyting algebras* HA *and the variety of closure algebras* CA *are closed under the operation $\bigotimes$ of tensor sum.*

One can illustrate the non-conservativity of the operation $\bigotimes$ on specially chosen Heyting algebras. For example, the so-called variety LC, i.e., the variety of Heyting algebras such that the identity $(a \to b) \vee (b \to a) = 1$ holds, is not closed under tensor sums. However, it is noteworthy that many interesting classes of algebras are closed under $\bigotimes$.

A.8.3 Proposition *The following classes of algebras are closed under tensor sums:*

(1) KC (*the class of Heyting algebras such that the law of weak excluded middle holds, i.e.,* $\neg a \vee \neg\neg a = 1$);
(2) SA (*the class of skeletal closure algebras*);
(3) BA (*the class of Boolean algebras*);
(4) MA (*the class of Halmos's monadic algebras*);
(5) Post (*the class of Post algebras*);
(6) Luk (*the class of Lukasiewicz algebras*).[12]

---

[12] *Editorial note*: For discussion of monadic algebras, see [13]. For discussion of Post algebras and Lukasiewicz algebra, see, e.g., [1].

Note that every Lukasiewicz algebra and every Post algebra is a Heyting lattice.

Let $X_k = \{0, 1\}$ (for all $k \in \omega$) be the two-point space with the discrete topology Ω_k. Define an order R_k on X_k as follows: $x R_k y \Leftrightarrow x = y$ or $x = 0$ and $y = 1$. Consider the product of hybrids $(X, \Omega, R) = \prod_{i \in I}(X_k, \Omega_k, R_k)$.

A.8.4 Proposition (X, Ω, R) *is a strict hybrid and the corresponding Heyting algebra H is a free distributive lattice on countably many generators. Moreover, the lattice H is isomorphic to the coproduct (in the category* DL*) of the three-element Heyting lattices H_k $(k \in \omega)$.*

Let (X_i, R_i) $(i = 1, 2)$ be hybrids such that (X_2, R_2) is a cluster, i.e., for all $x, y \in X_2$, $x R_2 y$ and $y R_2 x$. Let (X, R) be their product. Figuratively speaking, (X, R) is obtained from (X_1, R_1) by blowing every point $x \in X_1$ up into a cluster A_x of size X_2 (i.e., A_x and X_2 are of the same cardinality) and the order among clusters is the same as the order among points of X_1; in other words, the skeleton of the hybrid (X, R) coincides with (X_1, R_1). The blowup procedure has interesting applications; in particular, it features in the proof of the Dummett–Lemmon conjecture [5].

In conclusion, we note that the operation $\bigotimes$ of tensor sum induces the following operation on varieties (and hence on modal systems and superintuitionistic logics). If K_1, K_2 are varieties of Heyting algebras, then let $K = K_1 \bigotimes K_2$ be the smallest variety generated by the class of Heyting algebras of the form $H_1 \bigotimes H_2$, where $H_1 \in K_1$ and $H_2 \in K_2$.

(B) Alternation [8]. Consider another non-conservative operation, which we call the operation of **alternation**. Its definition is motivated by S. Jaskowski's Γ-operation.[13]

Let $(B_i, \mathbf{C}_i)$ be a closure algebra and H_i its skeleton, indexed by elements of a partially ordered set $(I, \leq)$.

The **alternation** $\#_{i \in I}(B_i, \mathbf{C}_i) = (B, \mathbf{C})$ of closure algebras $(B_i, \mathbf{C}_i)$ $(i \in I)$ is defined as follows:

(a) $B = \prod_{i \in I} B_i$ is the direct product of the Boolean algebras B_i,

(b) $(\mathbf{C} f)_i = \mathbf{C}_i f_i$ if $(\forall j)(i < j \Rightarrow f_j = 0)$ and $(\mathbf{C} f)_i = 1$ otherwise; here $f \in B = \prod_{i \in I} B_i$, and f_i is the projection of f onto the component B_i.

A.8.5 Proposition *The alternation* $\#_{i \in I}(B_i, \mathbf{C}_i)$ *of a family of closure algebras* $\{(B_i, \mathbf{C}_i) : i \in I\}$*, for a partially ordered set I, is a closure algebra.*

Note that if the partially ordered set $(I, \leq)$ is an antichain, i.e., $x \leq y \Leftrightarrow x = y$, then it is immediate from the definition that the alternation $\#_{i \in I}(B_i, \mathbf{C}_i)$ is the direct product $\prod_{i \in I}(B_i, \mathbf{C}_i)$ of the algebras $(B_i, \mathbf{C}_i)$ $(i \in I)$.

---

[13] *Editorial note*: See S. Jaśkowski, "Recherches sur le Système de la Logique Intuitioniste," *Actes du Congrès International de Philosophie Scientifique* 6: 58–61, 1936. An English translation is available in S. Jaśkowski, "Investigations into the System of Intuitionist Logic," *Studia Logica* 34(2): 117–120, 1975.

A definition of the operation dual to alternation is as follows. Here we restrict ourselves to a finite partially ordered set I of indices.

By definition, a hybrid (X, Ω, R) is the **ordered sum** of hybrids (X_i, Ω_i, R_i) $(i \in I)$ if:

(a) the space (X, Ω) is the topological sum of the spaces (X_i, Ω_i);
(b) for all $x, y \in X$, xRy iff either $x \in X_i$ and $y \in X_j$ $(i \leq j, i \neq j)$ or $x, y \in X_i$ and xR_iy (in other words, (X, R) is obtained by "positioning" the quasi-ordered sets (X_i, R_i) according to the partially ordered set $(I, \leq)$, preserving the original order R_i inside every (X_i, R_i)!).

Note that in the case of $(I, =)$, the quasi-ordered sets (X_i, R_i) are positioned in a "row" without any link between any of them.

A.8.6 Proposition *Let $(I, \leq)$ be a finite partially ordered set and (X_i, Ω_i, R_i) a hybrid for each $i \in I$. Then:*

(1) *the ordered sum (X, Ω, R) of these hybrids is a hybrid;*
(2) *if (X_i, Ω_i, R_i) is strict for each $i \in I$, then the hybrid (X, Ω, R) is strict.*

Using the Duality Theorem (Theorem 3.3.4), we can assume that the operation of alternation $\underset{i \in I}{\#} H_i$ is also defined for arbitrary Heyting algebras. In particular, if $(I, \leq) = (\{1, 2\}, \leq)$ is the two-element chain, then in our terminology Jaskowski's Γ-operation can be written as follows: $\Gamma(H_2) = H_1 \# H_2$, where H_1 is the two-element Boolean algebra and H_2 is an arbitrary Heyting algebra; or, passing to the corresponding hybrids (X_i, Ω_i, R_i) $(i = 1, 2)$, we obtain their ordered sum; here (X_2, Ω_2, R_2) is an arbitrary hybrid and (X_1, Ω_1, R_1) is the one-point hybrid "situated under" (X_2, Ω_2, R_2).

Even this very special case of the operation of alternation demonstrates its equational non-conservativity: recall the known method of a characteristic sequence of Heyting algebras starting from the two-element Boolean algebra by iterating the operations of direct product and Jaskowski's Γ-operation.

A.9 Boolean Cascades

The content of this section is based on [10]. In Part II of the book, certain special classes of Heyting algebras and closure algebras are analyzed in terms of hybrids:

(1) Stone algebras;
(2) Horn's LC-algebras;
(3) Halmos's monadic algebras;
(4) Closure algebras corresponding to McKinsey's interesting modal system **S4.1** ("Lions in Alaska"[14]);

[14] *Editorial note*: See J.C.C. McKinsey, "On the syntactical construction of systems of modal logic," *The Journal of Symbolic Logic*, Vol. 10, No. 3, 1945, pp. 83–94.

(5) A complete description of the free cyclic algebra of Makinson's variety D^*;
(6) Cascade Heyting algebras.

We comment only on this last item because the indicated class of Heyting algebras corresponds to a superintuitionistic logic which has not been considered previously in the literature.

Let **PWL** (called the **logic of Peirce's weak law**) be the calculus obtained from the intuitionistic propositional calculus by postulating the formula

$$(q \to p) \vee \big(((p \to q) \to p) \to p\big)$$

as an additional axiom (which we call **Peirce's weak law**).

Let CHA be the class of Heyting algebras satisfying Peirce's weak law

$$(b \to a) \vee \big(((a \to b) \to a) \to a\big) = 1$$

(cf. Peirce's law in Proposition 2.1.5). We call the algebras in this class **cascade Heyting algebras**.

The **cascade join** of a finite sequence of Heyting algebras H_i $(1 \leq i \leq k; k \in \omega)$ is, by definition, the Heyting algebra H that is the union of the sublattices H_i' isomorphic to the lattices H_i $(1 \leq i \leq k)$ such that the intersection of H_i' and H_{i+1}' $(1 \leq i < k)$ contains exactly one element that is the least element of H_{i+1}' and the greatest element of H_i'. If a Heyting algebra H is isomorphic to the cascade join of a finite sequence of Boolean algebras, then we call H a **Boolean cascade**.

A.9.1 Proposition

(1) *The variety* CHA *of cascade Heyting algebras is generated by the class of finite Boolean cascades and is locally finite.*
(2) *The smallest variety of closure algebras generated by the skeletal closure algebras* $B(H)$*, where* $H \in \mathsf{CHA}$*, is the variety of Grzegorczyk algebras satisfying the additional identity*

$$\mathbf{C}(a - \mathbf{C}b) \wedge \mathbf{C}\big(a - \mathbf{C}\big(\mathbf{C}a - \mathbf{C}(\mathbf{C}b - \mathbf{C}a)\big)\big) = 0.$$

This variety is hereditarily finitely approximable.

Let (X, R) be a partially ordered set. We call the following condition the **three point condition**: for any three distinct points $x, y, z \in X$, if x and y are incomparable, then xRz implies yRz.

A.9.2 Proposition *Let* H *be a subdirectly irreducible Heyting algebra and* X *the corresponding hybrid. Then the following are equivalent:*

(1) H *is a cascade algebra;*
(2) *the partially ordered set* (X, R) *satisfies the three point condition.*

A.10 Cantor's Scattered Spaces

In this section we recall a few important and fruitful century-old ideas due to Cantor and, with pleasure, discover their close connection with Grzegorczyk's and Gödel's modal systems.

(1) The problem of representation of functions by trigonometric series was a focus of attention for many mathematicians when (100 years ago) George Cantor, trying to extend the uniqueness of representation to a wider class of functions and not yet having discovered set theory, introduced the concept of **derivative** of infinite order. This concept was not only the starting point of research in point-set topology, but it also lead Cantor to the construction of transfinite ordinal numbers and later to the discovery of set theory. The celebrated diagonal construction of Cantor and the concept of **scattered space** are also of that period. We start by recalling the main definitions.

A point x of a topological space X is called a **limit point** of a set A if for every neighborhood U_x of x we have $(A \cap U_x) - \{x\} \neq \varnothing$. We denote the set of all limit points of A by dA. The set dA is called the **derivative** of A. The derivatives of order α are defined by transfinite induction: $A^0 = A$, $A^{\alpha+1} = dA^\alpha$ and $A^\lambda = \bigcap\{A^\alpha : \alpha < \lambda\}$ for a limit ordinal λ.

It is known that in introducing ordinal numbers Cantor was motivated by the following concept. A topological space X is said to be **scattered** if it does not contain a nonempty subset that is dense-in-itself (Cantor). Recall that a space X is scattered if and only if there exists an ordinal α such that $X^\alpha = \varnothing$ and that every ordinal is a scattered space.[15]

(2) It has been recently discovered that the modal systems **G** of Gödel and **Grz** of Grzegorczyk, which are closely related to each other, can be applied to the analysis of mathematical results related to Peano Arithmetic.[16] Both systems allow adequate arithmetical interpretations: for the former system, interpret the modality $\Box p$ as $\mathrm{Bew}(p)$, where Bew is the predicate of provability in Peano Arithmetic (R. Solovay); for the latter system, interpret $\Box p$ as $\mathrm{Bew}(p)\ \&\ p$ (A.V. Kuznetsov, R. Goldblatt, G. Boolos).

Now we turn to the connections between (1) and (2).

A.10.1 Proposition *Grzegorczyk's modal system* **Grz** (*resp. Gödel's modal system* **G**) *is characterized by the class of Cantor's scattered spaces if the operator of possibility* $\Diamond$ *is interpreted as the operator of topological closure* (*resp. as the operator of topological derivative*).

The proof of this statement is based on the following *new* axiomatic definition of Cantor's scattered spaces, which, in our point of view, is of independent interest.

---

[15]*Editorial note*: See, e.g., Sect. 8.5 of Z. Semadeni, *Banach Spaces of Continuous Functions*, Vol. 1 (PWN – Polish Scientific Publishers), 1971.

[16]*Editorial note*: The system **G** is the smallest normal modal logic containing the **Löb axiom** $\Box(\Box p \to p) \to \Box p$ and is now called the Gödel–Löb logic **GL**. See, e.g., G. Boolos, *The Logic of Provability* (Cambridge University Press), 1993.

A.10.2 Proposition (An axiomatic definition of Cantor's scattered spaces) *The class of scattered spaces is exactly the class of sets X equipped with an operator d satisfying the following conditions:*

(1) $d\varnothing = \varnothing$;
(2) $d(A \cup B) = dA \cup dB$;
(3) $dA = d(A - dA)$ (*Löb's formula*!).

In other words, the class of scattered spaces coincides with the class of topological spaces X such that the operator of derivative on X satisfies "Löb's identity" $dA = d(A - dA)$ (for all sets A). We conclude by referring the reader to [9] for more details.

A.11 Symmetric Heyting algebras

A number of authors (G. Moisil, A.V. Kuznetsov, A. Muchnik, S. Rauszer, L. Iturioz) have suggested "symmetric" formulations of intuitionistic propositional calculus **SIn**. In this calculus, for each connective $\&$, $\vee$, $\rightarrow$, $\neg$, there is a dual one $\vee$, $\&$, $\leftarrow$, $\sim$, and the duality principle of classical logic is restored. On the other hand, McKinsey and Tarski [15] had already paid special attention to the difference between Boolean lattices and Heyting lattices due to the failure of the duality principle in Heyting lattices. On p. 141 of that work, the authors write: "The problem of dualism in Brouwerian algebra is not yet clear." McKinsey and Tarski indicated that the class HL of Heyting lattices is very different from both the class DL of distributive lattices and the class BL of Boolean lattices (which is a subclass of HL) because even the weak duality principle fails in HL. In the quoted paper, they introduced the concept of bi-Heyting algebra (this concept, as the authors indicate, had been considered by T. Skolem as early as 1919).

Let $(H, \leq)$ be a bounded distributive lattice. Put the lattice H "upside-down," i.e., re-order the set H as follows: $a \leq^{\circ} b \Leftrightarrow b \leq a$. It is easy to see that $(H, \leq^{\circ})$ is a bounded distributive lattice. So the class of distributive lattices is **stable** with respect to the operation of "turning things upside-down." The class BL of Boolean lattices has the property of **strict stability**. Namely, if $(B, \leq) \in \mathsf{BL}$, then $(B, \leq^{\circ})$ not only remains in BL but is even isomorphic to the original one, i.e., $(B, \leq) \simeq (B, \leq^{\circ})$. On the other hand, the class of Heyting lattices is not stable: it is known that the family Ω of all open sets of any topological space, ordered by inclusion $\subseteq$, is a Heyting lattice; however, the lattice $(\Omega, \subseteq^{\circ})$ (which is obviously isomorphic to the lattice of closed subsets of X) is, in general, not a Heyting lattice.

We call a Heyting lattice $(H, \leq)$ **symmetric** (resp. **strictly symmetric**) if $(H, \leq^{\circ})$ is a Heyting lattice (resp. a Heyting lattice isomorphic to the original one $(H, \leq)$).

A.11.1 Proposition [6] *Let H be a Heyting lattice and* (X, R) *its hybrid. The following are equivalent*:

(1) *H is symmetric (resp. strictly symmetric)*;
(2) $(X, \check{R})$ *is a hybrid (resp. a hybrid isomorphic to the hybrid* (X, R)*), where* $x\check{R}y \Leftrightarrow yRx$ *for all* $x, y \in X$.

References

1. Balbes, R., & Dwinger, Ph. (1974). *Distributive lattices*. Columbia, MO: University of Missouri Press.
2. Birkhoff, G. (1967). *Lattice theory* (3rd ed., Vol. XXV). Providence: American Mathematical Society Colloquium Publications, American Mathematical Society.
3. Blok, W. J. (1976). Varieties of interior algebras. Ph.D. thesis, University of Amsterdam.
4. Esakia, L. L. (1976). On modal "companions" of superintuitionistic logics (Russian). In *VII Soviet Symposium on Logic (Kiev, 1976)* (pp. 135–136).
5. Esakia, L. L. (1976). On the Dummett-Lemmon conjecture (Russian). In *IV Soviet Conference in Mathematical Logic (Chisinau, 1976)* (p. 160).
6. Esakia, L. L. (1978). Semantical analysis of bimodal systems (Russian). In *Logic, Semantics and Methodology, Metsniereba, Tbilisi* (pp. 87–99).
7. Esakia, L. L. (1979). Duality for almost direct products of Boolean, Heyting and closure algebras. In *VI International Congress of Logic, Methodology and Philosophy of Science, West Germany* (pp. 103–106).
8. Esakia, L. L. (1979). On the theory of modal and superintuitionistic systems (Russian). In *Logical Inference*. Moscow: Nauka (pp. 147–172).
9. Esakia, L. L. (1981). Diagonal constructions, Löb's formula, and Cantor's scattered spaces (Russian). In *Studies in Logic and Semantics, Metsniereba, Tbilisi* (pp. 128–143).
10. Esakia, L. L. (1983). On a locally finite variety of Heyting algebras (Russian). In *XVII Soviet Algebraic Conference. Part II, Minsk* (pp. 280–281).
11. Esakia, L. L., & Grigolia, R. (1981). The variety of Heyting algebras is balanced (Russian). In *XVI Soviet Algebraic Conference. Part II, Leningrad* (pp. 37–38).
12. Grätzer, G. (1978). *General lattice theory*. Basel: Birkhäuser Verlag.
13. Halmos, P. R. (1962). *Algebraic logic* New York: Chelsea Publishing Co.
14. McKinsey, J. C. C., & Tarski, A. (1944). The algebra of topology. *Annuals of Mathematics*, 2(45), 141–191.
15. McKinsey, J. C. C., & Tarski, A. (1946). On closed elements in closure algebras. *Annals of Mathematics*, *2*(47), 122–162.
16. Rasiowa, H., & Sikorski, R. (1963). *The mathematics of metamathematics*. Monografie Matematyczne, Tom 41. Warsaw: Panstwowe Wydawnictwo Naukowe.

Index

G. Bezhanishvili and W. H. Holliday (eds.), *Heyting Algebras*,
Trends in Logic 50, https://doi.org/10.1007/978-3-030-12096-2

www.ingramcontent.com/pod-product-compliance
Lightning Source LLC
Chambersburg PA
CBHW070759180625
28394CB00003BA/59
9783030120955